Top 20 Essential Skills
for Imagery and Remote Sensing

TOP 20

ESSENTIAL SKILLS FOR IMAGERY AND REMOTE SENSING

Thomas Humber

Jeff Liedtke

Esri Press

Redlands, California

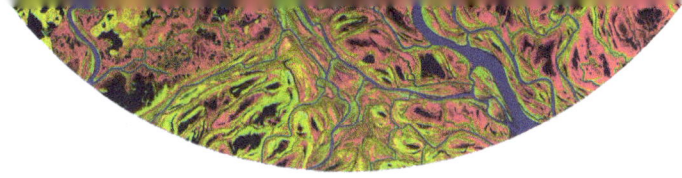

CONTENTS

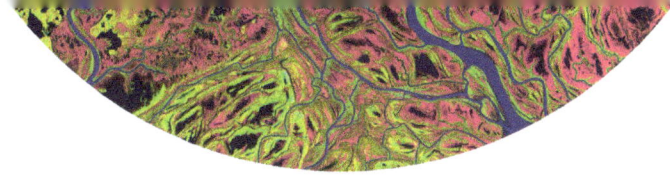

FOREWORD

For much of the last 30 years, we have watched the pace of innovation in the collection and processing of remotely sensed imagery accelerate exponentially. From spacecraft to aircraft to drones and mobile 360 data collection platforms, the volume, variety, and velocity at which data is now being collected and curated is staggering. In the early days, those of us working with this imagery could only imagine the possibilities. We would be ecstatic to get a handful of new scenes every few months to use in developing and testing our processing tools and algorithms. Today, we have a constant stream of imagery coming to us, and thanks to the scalability of cloud infrastructure, we now have the storage and processing power to ingest and use the insights from all this data—now in near real time and inching closer to real time with each passing day. These collection systems, and the data they produce, are an indispensable component of a network of devices that monitor what is happening on the earth and bring us closer to realizing what we at Esri® have long envisioned as a central nervous system for the planet.

In today's increasingly interconnected and data-rich world, the ability to observe and understand our planet from above is no longer a luxury—it's a necessity. Remote sensing has evolved from a niche scientific discipline into a cornerstone of modern geospatial intelligence, empowering professionals across sectors and industries to make informed decisions with unprecedented precision, timeliness, and scale.

In the early days of the discipline, those of us in the remote sensing fields never really worked directly with our colleagues in the geographic information system (GIS) domain. Now the two fields are inseparable, requiring an integrated platform and toolbox for managing, visualizing, processing, and analyzing every imaginable type of remotely sensed imagery within a dynamic mapping context, driving increasingly complex spatial analytics. For the last 10 years, this has also fueled considerable innovation in the application of computer vision–based artificial intelligence (AI) models for feature extraction, change detection, and even foundational map production. The leading edge of this work is now shifting toward AI-powered automated workflows that will soon drive multiagent systems tasked with monitoring supply chains, assessing risks to assets, detecting changes in forests and land cover, overseeing construction sites, responding to natural disasters, and supporting efforts to combat the impacts of climate change.

As data proliferates, imagery use cases and applications continue to expand. This expansion is also being characterized by a growing shift of uses of remotely sensed data from governments to commercial organizations. Some in the field estimate that by 2030 private organizations will account for more than 50% of the use of remotely sensed imagery. In the private sector, it is driving innovation in agriculture, energy, transportation, insurance, and environmental consulting, enabling companies to optimize operations, reduce risk, and unlock new value from spatial data.

Before we get too far ahead of ourselves getting excited about the possibilities of working with imagery across the ArcGIS® platform, we must also recognize that we will need to get more people on board with our vision. We must start at the beginning and expand the community of remote sensing professionals, beginning with the broader community of GIS professionals. Many are finding their way there naturally, and others are just becoming aware of the power, and the necessity, of making imagery a foundational part of their GIS work. This is why we have this book and designed it to meet the needs of both seasoned practitioners and those just beginning their journey. It offers a clear, practical foundation in remote sensing principles, technologies, and applications, with real-world examples that illustrate how satellite and aerial imagery is transforming decision-making across industries.

For students and emerging professionals, this text serves as both a guide and an invitation. Remote sensing is not just about pixels and platforms. It's about seeing the world differently, asking better questions, and solving complex problems with clarity and creativity. Whether you're studying environmental science, urban planning, or data analytics, understanding remote sensing will expand your capabilities and open doors to impactful careers.

As someone who has worked at the intersection of GIS and remote sensing, I've seen firsthand how these tools can illuminate patterns, reveal change, and inspire action. This book captures that potential and equips readers with the knowledge to harness it. It is a timely and essential resource for anyone committed to using geospatial technology to better understand—and improve—the world around us. Come and join us on this important geospatial journey.

Richard Cooke
Corporate Director
Esri

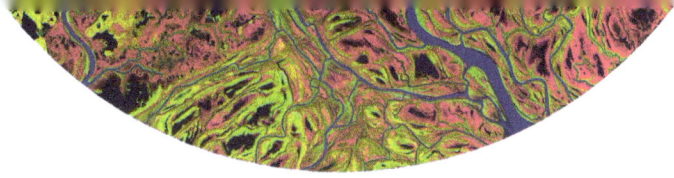

ACKNOWLEDGMENTS

This book was made possible through collaboration with several Esri domain experts, who generously shared their knowledge by authoring chapters in their areas of specialty. Their expertise and dedication not only strengthen the book but also highlight the spirit of teamwork and shared learning at the heart of this work. We acknowledge and thank the contributing authors:

- **Hong Xu:** Hong is a principal software product engineer on Esri's Imagery and Remote Sensing team and has played key roles in the development and leadership of various software products related to imagery and data science. Currently, Hong's areas of focus include time series image analysis, multidimensional raster, hyperspectral imagery, and altimetry data. Hong is also a frequent *ArcGIS Blog* contributor on imagery and remote sensing topics.

- **Elizabeth Ashley Menezes:** Elizabeth is a geophysicist and radar specialist on the Imagery and Remote Sensing team, mapping the planet with satellites one pixel at a time. Elizabeth is passionate about science communication and dedicated to making learning accessible through thoughtful, intentional design. Elizabeth believes science should be beautiful, understandable, and shared widely.

- **Jeff Swain:** Jeff is a product engineer on the Imagery and Remote Sensing team and focuses on solving problems through imagery and shared understanding, always driven by a passion to make things work better. Jeff says, "I'm grateful for the creative minds I've met along the way—collaboration has shown me that the best solutions are often born from many voices working together."

- **Pavan Yadav:** Pavan is a senior software product engineer on the Imagery and Remote Sensing team. Pavan uses AI to extract valuable insights from imagery data and works to contribute to the development of geospatial AI (GeoAI).

- **Simon Woo:** Simon is a product engineer on the Imagery and Remote Sensing team. Over the last 20 years, Simon has worked in areas such as image visualization, georeferencing, Pixel Editor, and SAR.

- **Christopher Patterson:** Chris serves as a principal product engineer on the Imagery and Remote Sensing team, specializing in the efficient processing of imagery within ArcGIS Pro. Chris's areas of expertise encompass Reality mapping, ortho mapping, and stereoscopic mapping. Chris is committed to supporting others in effectively using these technologies.
- **Tracy Toutant:** Tracy is a senior product manager on Esri's National Imagery team and leads the development of geospatial video in both ArcGIS Pro and ArcGIS Video Server. Tracy specializes in the fusion of imagery and GIS data and its application in situation awareness and decision support.
- **Richard Cooke:** Richard is Esri Director of Global Business Development. Richard works with customers across government and private enterprise to deliver value and improve business and policy outcomes through the application of GIS technology. Richard has more than 30 years of experience in the fields of computer vision, remote sensing, and geospatial and location analytics.

We acknowledge and thank the Esri Press team—Craig Carpenter, Maryam Mafuri, Carolyn Schatz, David Oberman, and Victoria Roberts—for their support and dedication in planning, testing, editing, and designing this book. Additional thanks to Elizabeth Ashley Menezes for providing many graphics and ideas to help represent and simplify conceptual topics.

Thomas: For Leslie, everything I do…

Jeff: To my wife, for her steadfast support and encouragement, and my two young sons, who sacrificed several impromptu wrestling matches for this work.

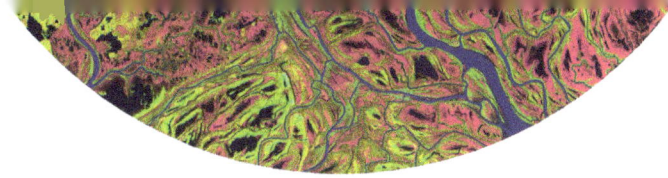

INTRODUCTION

Today, imagery is everywhere, even on our phones and mobile devices. We love that! When we began our careers decades ago, analyzing and even viewing imagery required specialized software, hardware (array processors), high-power computers, and sometimes unique access to data. Projects and work were hindered by significant technical limitations. No more. Today, advancements in both technology and accessibility have enabled us to maximize the potential of remotely sensed imagery. GIS, in particular, has greatly benefited from incorporating imagery into analysis and processing workflows.

Traditionally, the discipline of image processing and remote sensing evolved alongside, but separately from, the development of GIS technology, with limited analytical interaction. Even among scientists, analysts, practitioners, and professionals—either focusing on GIS or imagery and remote sensing—the two fields seemed related but separate. ArcGIS technology has always provided the capability to view rasters and imagery, but it was traditionally used only as a backdrop to support vector-based data. Over the past decade, however, Esri has embedded remote sensing, image processing, and GIS analysis into ArcGIS.

One of ArcGIS's most important strengths is its ability to fully integrate image processing and remote sensing with traditional GIS capabilities. By combining imagery and GIS vector data within a single platform, you can seamlessly conduct both raster and vector analyses without the need for extensive conversions or workarounds. This capability streamlines workflows, ensures consistency in processes and projects, including elements of coordinate systems and metadata, and strengthens overall data management. More significantly, this integration enriches geospatial modeling and analysis, supports enterprise-level interoperability, and enables the generation of advanced insights and solutions.

Beyond these benefits, Esri has also emerged as a leading provider of cutting-edge remote sensing and image-processing capabilities. What do we mean by cutting edge image processing? Over the years, Esri has developed advanced capabilities for managing and analyzing hyperspectral imagery, synthetic aperture radar (SAR) data, scientific multidimensional data, and rigorous photogrammetric processing and has incorporated next-generation analytic capabilities, such as deep learning and machine learning.

Early multispectral analysis was largely limited to two dimensions. The introduction of terrain data and digital elevation models (DEMs) extended this into the realm of 3D, which was later expanded with multidimensional datasets incorporating height, depth, and time. Today, the immense "fire hose" of imagery and data—often collected from multiple sources over the same location—has elevated time to a central dimension of analysis.

Time series analysis now allows us to uncover trends and explore cause-and-effect relationships. This deeper temporal understanding is crucial for solving complex challenges, from assessing environmental change to addressing the multifaceted impacts of climate change across physical, ecological, and societal domains. It also unlocks a powerful new application: the ability to monitor situations and phenomena. Once limited or even impossible, monitoring applications have now become a core capability. These advancements rely on robust data models, such as data cubes, which provide the architectural foundation needed to manage, analyze, and interpret this new level of multidimensional information.

Our goal with *Top 20 Essential Skills for Imagery and Remote Sensing* is to continue to bridge the worlds of GIS and remote sensing. The content is organized thematically to guide readers through the fundamentals and preparation, data exploration, and advanced analysis. To enrich this journey, we have drawn on the expertise of Esri colleagues and leading subject matter experts. The lessons in the book will equip users to address real-world workflows and are designed to help readers apply image processing and remote sensing skills in their day-to-day work. To these ends, the final chapter culminates in a decision support workflow, illustrating where and how imagery can best be applied while tying these methods to the skills and concepts introduced earlier.

Finally, for readers seeking a deeper scientific grounding and conceptual underpinnings to the topics and workflows presented here, we recommend *Imagery and GIS: Best Practices for Extracting Information from Imagery* by Kass Green, Russell G. Congalton, and Mark Tukman (Esri Press, 2017). Used together, these two works provide a solid foundation for understanding and applying imagery and GIS technologies in ways that may unlock synergistic solutions to some of today's most urgent challenges.

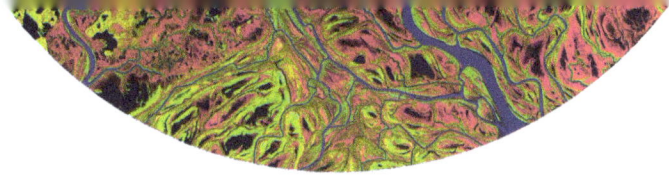

HOW TO USE THIS BOOK

About this book

Top 20 Essential Skills for Imagery and Remote Sensing has been tested for compatibility with ArcGIS® Pro 3.5. This book is designed for beginners to intermediate users. No prior experience with imagery is required. The tutorials in each chapter demonstrate a workflow in a hands-on environment and should take about 45 minutes each to complete.

Software requirements and licensing

To perform the tutorials in this book, you'll need the following: ArcGIS Pro 3.5 or higher and a web browser to access ArcGIS Online. A user type of Professional or Professional Plus is needed for the image analysis sections.

Two chapters—chapter 7, "Creating Photogrammetric Products," and chapter 13, "Using Deep Learning for Object Detection and Classification"—require additional setups. Setup instructions are included in these chapters.

Earlier software versions may not be fully compatible with the tutorial data and may not operate as described in the steps.

Hardware requirements for ArcGIS Pro are available at links.esri.com/SysReqs.

Information on software trial options, as well as Personal Use and Student Use licensing, can be found at esri.com.

Downloading the tutorial data

At the beginning of each chapter, a link to that chapter's data is given along with setup instructions. In some cases, the datasets are large. Each chapter's dataset is provided as a zip file, which should be unzipped and stored in an enclosing folder, Top20Imagery, that you will need to create. For example, the folder structure for the second chapter should be as follows: C:\Top20Imagery\Top20Imagery_02.

Tutorial data that accompanies this book is covered by a license agreement that stipulates the terms of use. You can view the agreement at links.esri.com/License-Agreement.

Resources, feedback, and updates

For readers seeking deeper scientific grounding, many of the concepts presented here build on subjects discussed in more depth in *Imagery and GIS: Best Practices for Extracting Information from Imagery* by Kass Green, Russell G. Congalton, and Mark Tukman (Esri Press, 2017). These two works together provide a solid curriculum for understanding image processing and remote sensing fundamentals, which are applied using the tutorials in this book.

The ArcGIS Pro Help documentation provides comprehensive descriptions of software concepts and tools at links.esri.com/Help.

Feedback, updates, and collaboration options are available at Esri Community, the global community of Esri users. Post any questions about this book at links.esri.com /EsriPressCommunity.

Visit the book's web page at links.esri.com/Imagery20.

Facing page

The image at the beginning of each part of the book is the Yukon River Delta in southwestern Alaska, one of the largest river deltas in the world, and is currently protected as part of the Yukon Delta National Wildlife Refuge. The image, captured by the Landsat 7 satellite in 2002, is part of the USGS "Earth as Art" collection. The band combination is 7 (SWIR), 4 (NIR), and 1 (Blue). Courtesy of the Earth Resources Observation and Science (EROS) Center.

PART 1

Fundamentals and preparing your data

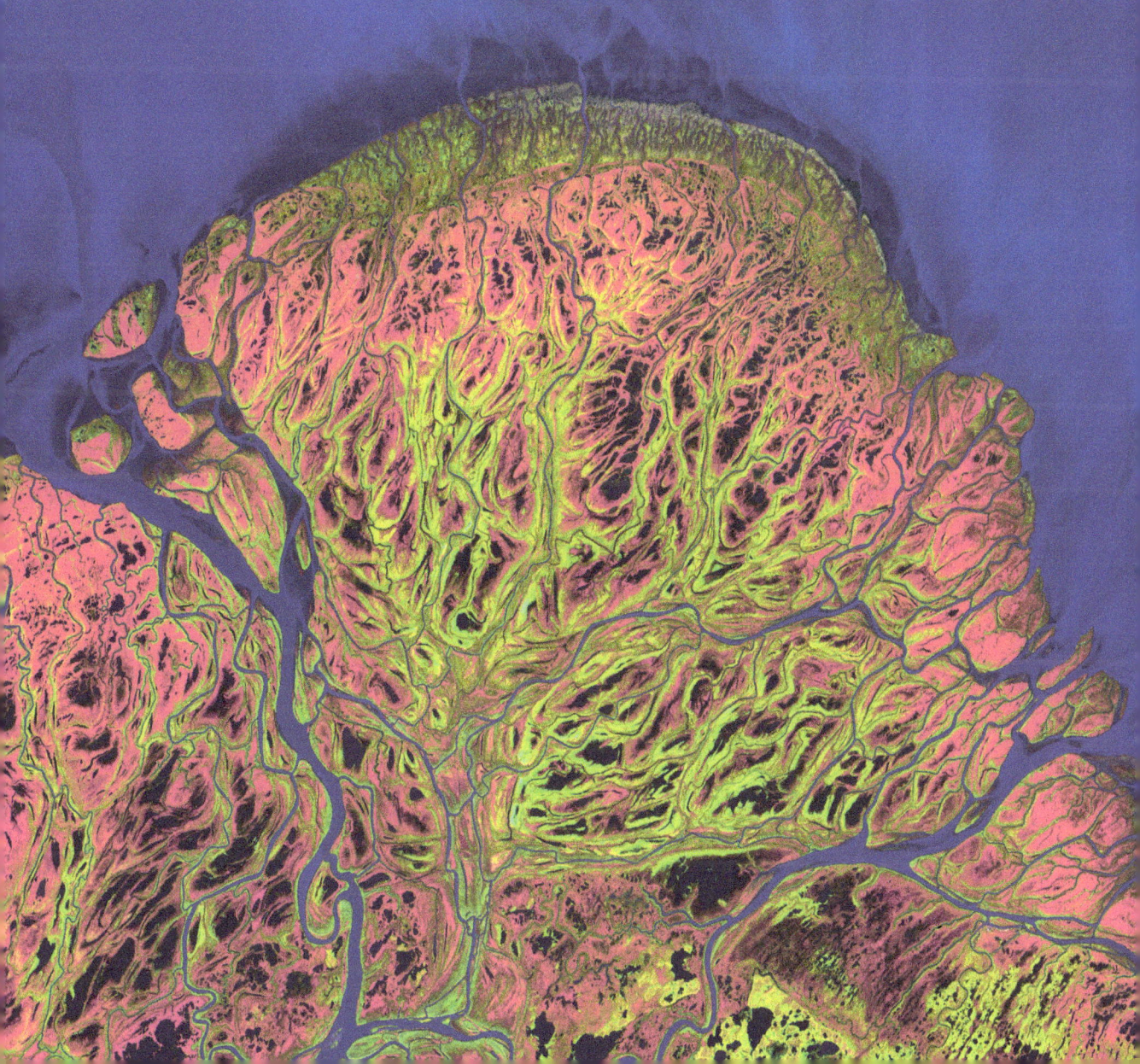

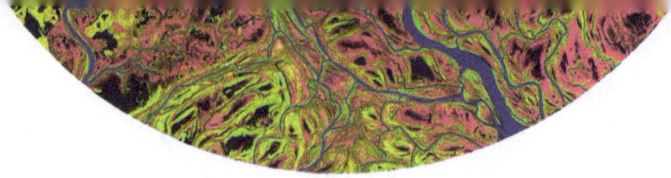

CHAPTER 1

Understanding imagery and remote sensing

Objectives

- Explore concepts of image processing and remote sensing used in ArcGIS.
- Learn a working vocabulary for imagery in ArcGIS.
- Get an introduction to ArcGIS Pro imagery information models.

Introduction

The adage "If I can see it, I can understand it" defines the power of imagery. Imagery gives an all-important geospatial context for better understanding of issues and solutions. The information content in imagery is unparalleled and supports a wide variety of applications in many industries.

ArcGIS Pro provides many tools and capabilities for visualizing, managing, processing, and analyzing imagery and raster data. The ability to use, render, and understand imagery within the context of GIS continues to be a competency that analysts require. As you learn these essential skills in this book, you will also learn how imagery can enhance the context, accuracy, and effectiveness of your analysis in several important ways.

- **Real-world representation:** Imagery provides a georeferenced visual snapshot of the earth's surface, accurately capturing features that can be derived through visualization or analysis, such as land cover or land use types, topography, infrastructure, and environmental conditions at specific points in time. This data serves as foundational base layers in GIS, enabling precise spatial contexts for mapping, analysis, and feature extraction. Imagery's unique capability to combine spectral, spatial, and temporal characteristics also supports the interpretation of surface materials and landscape dynamics, enhancing the realism and analytic depth of GIS projects.

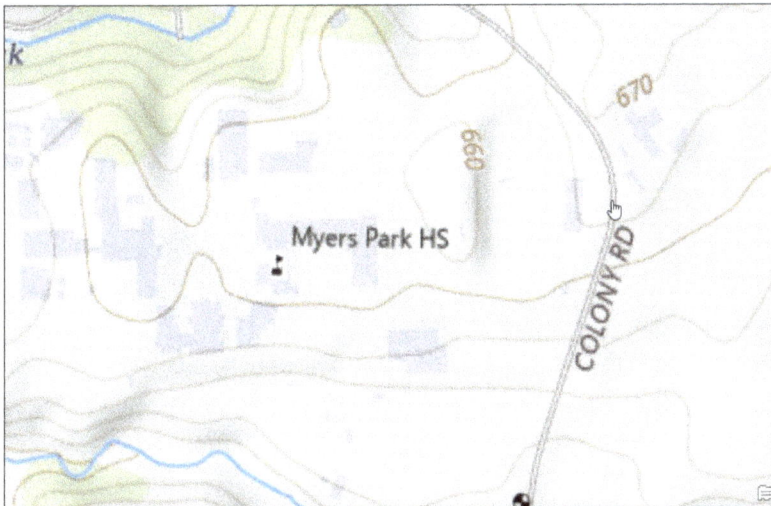

Figure 1-1. An image provides more contextual information than a symbolic map.

- **Timely and dynamic data:** Imagery often represents the most current condition of a location and can thus enhance and augment critical time-sensitive situations. Current imagery improves visual understanding of a situation, making it easier to communicate risks or conditions to both technical and nontechnical audiences. Frequently updated imagery enables monitoring of change over time, such as urban growth, deforestation, flooding, or varying agricultural activity, making it essential for real-time or time series analysis.

Figure 1-2. An image provides the most up-to-date information and is invaluable for emergency response and mitigation, such as this tornado track.

- **Advanced geospatial analysis:** Remote sensing techniques, such as classification, vegetation indices, or object detection, can be used on imagery to extract valuable insights, including land use types, soil moisture, or vegetation health—information that vector data alone cannot provide.

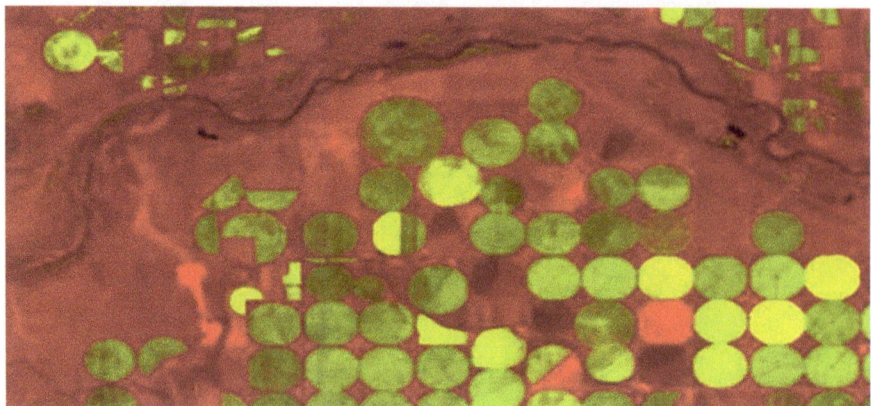

Figure 1-3. Remote sensing analysis reveals crop health in a Landsat 8 image.

- **Enhanced decision-making:** By adding visual and geospatial context, imagery helps decision-makers understand spatial relationships, assess conditions on the ground, and make informed choices in planning, coordinating field teams, enhancing resource management, and helping to facilitate effective emergency response.
- **Integration with GIS data:** Imagery integrates seamlessly with GIS layers (points, lines, polygons) to validate, enrich, and guide GIS analysis, allowing you to digitize features, verify field data, and detect errors in existing datasets.

ArcGIS advantages

ArcGIS offers numerous benefits for image processing and remote sensing applications, many of which are discussed throughout this book. These advantages are largely driven by two fundamental principles built into the ArcGIS framework:

- Seamless integration of image processing and GIS functionality
- Incorporation of metadata into fundamental data models

Integration of imagery and GIS

The integration of image processing and remote sensing with traditional vector-based GIS is a unique strength of ArcGIS, offering numerous advantages. This seamless combination of imagery and GIS layers within a single platform allows you to efficiently perform both raster and vector analysis, streamlining workflows and minimizing data conversion errors. It ensures consistency in coordinate systems, metadata handling, and overall data management. For example, after correcting a collection of imagery using rigorous photogrammetric techniques, you can immediately use the generated products to update existing land base layers in your GIS without conversion to other coordinate systems or types of data.

Additionally, this integration provides crucial context and information for geospatial modeling and analysis, promotes interoperability across an enterprise, and facilitates the generation of advanced insights and solutions.

In short, the integrated capabilities of ArcGIS remove barriers between raster and vector analysis, making it a powerful, flexible, and efficient platform for analysis and spatial decision-making.

Incorporation of metadata into fundamental data models

Image source metadata enables the analysis and management of imagery by providing critical information, such as sensor type, acquisition date and conditions, spatial resolution, and processing history. It ensures proper scientific processing, drives automated workflows, and supports efficient organization and discovery of appropriate image collections. By preserving data lineage and consistency, metadata transforms raw imagery into reliable, actionable information for GIS and remote sensing applications. This metadata also allows for better filtering, such as selecting images by cloud cover or time range, and supports interoperability with standardized formats. Ultimately, it improves decision-making, reduces errors, and allows scalable, repeatable image analysis across diverse projects.

The disciplines of GIS and image processing and analysis have traditionally evolved concurrently but separately, with little overlap, over decades. ArcGIS—and other, third-party GIS apps—have always had the capability to handle rasters and imagery, usually as a backdrop to GIS layers. Consequently, image processing has been typically limited to image enhancement, map reprojection, and tools to help register imagery to GIS layers. As a result, GIS professionals are generally unfamiliar with image processing and remote sensing techniques and methods. ArcGIS has transcended these limitations with full-functioning image processing, management of large raster and image collections, and analysis using advanced remote sensing techniques.

The purpose of this chapter is to introduce you to the characteristics of imagery and rasters, discuss why they are important, and explain how imagery is integrated within ArcGIS. The goal of this orientation is to prepare you for success and to get the most benefit from using the upcoming tutorials. And to help establish an aptitude for a fascinating and rewarding journey in the world of image processing and remote sensing going forward.

What is imagery and remote sensing?

Images are a form of raster data that represents measurements of reflected or emitted electromagnetic energy captured by sensors mounted on drones, airplanes, or satellites. Other types of raster data include scientific measurements of various properties and variables collected at specific locations. These multidimensional rasters contain information, such as temperature or salinity at different water depths, elevation models, and seismic surveys.

Remote sensing involves extracting meaningful information from imagery using scientific methodologies, phenomenology, physical and environmental factors, and image processing techniques. These processes identify and extract types of information about features of interest, such as vegetation type and health, urban development patterns, and trends in detected objects. For example, remote sensing analysis considers vegetation growth cycles that elicit different spectral response characteristics. This analysis is then coupled with other data containing physical attributes, such as slope steepness and direction, precipitation and temperature, sun angle and intensity, stress factors, neighborhood and proximity factors, seasonality, physical and cultural geography, object shape and attributes, and other phenomena, and used to extract specific, meaningful information. The result of this comprehensive approach can then be used in subsequent analyses, decision support, or other decision-making.

Remote sensing embodies the principles of The Science of Where®. ArcGIS technology incorporates scientific and image processing principles through its tools and wizards that guide users through complex tasks to enable the extraction of meaningful spatial and spectral information from imagery to address general and specific projects and applications. Although these processes are standardized and optimized through well-known and well-researched advanced image processing techniques, workflows and results can vary based on specific scientific remote sensing methods.

Imagery and remote sensing provide a range of functions, tools, and capabilities to do the following:

- Correct, calibrate, and standardize imagery for workflow integration
- Perform photogrammetric corrections on remotely sensed imagery from drones, aircraft, and satellites
- Conduct image interpretation, processing, and analysis
- Perform multispectral, multidimensional, and hyperspectral image analysis and processing
- Create information products from imagery
- Implement advanced raster and image analysis workflows using machine learning, artificial intelligence (GeoAI), and feature extraction methodologies

ArcGIS Pro imagery information models

Before 2010, imagery in ArcGIS was used primarily as a backdrop to overlay GIS layers and information. Image analysis was limited to visualization and some traditional tools, such as ISO cluster and maximum likelihood classification. The introduction of the mosaic dataset data model transformed image analysis and management capabilities and helped propel ArcGIS to become one of the most popular, capable, and robust image processing and remote sensing analysis packages available today, and into the future.

The data and information models for imagery and rasters are fully integrated into ArcGIS and take advantage of metadata associated with imagery. This standard metadata is used and improved to create meta information, which determines how imagery is displayed, enhanced, and analyzed. For example, meta information drives how mosaic datasets are built, displayed, and managed; how image time series are displayed, analyzed, and graphed; and how photogrammetric correction is applied and products generated.

The imagery information models comprise six key components:

- Raster dataset
- Raster type
- Raster product
- Raster functions
- Mosaic dataset
- Image services

These fundamental components are addressed in turn.

Raster dataset

The raster dataset is the primary information model component, representing a basic image with basic behavior. Its role is to read and write image pixels and metadata.

Characteristics of raster datasets include the following:

- 1 or N bands
- 1–64 bits per pixel, per band
- Compressed or uncompressed
- More than 90 image and raster formats supported, with no need to convert

Raster type

The raster type is a fundamental information model component, representing the intelligent logic for a particular sensor or image product coming from a vendor. Its role is to provide the following capabilities:

- Define pixel storage and metadata schema
- Define the rules for ingesting imagery into ArcGIS
- Define any default processing chains
- Define any georeferencing (sensor model plus parameters)
- Use sensor and format specification for proper processing and analysis
- Support the construction of a mosaic dataset

Raster product

The raster product uses ArcGIS user interface shortcuts to enable standard band combinations and processing chains, allowing you to focus on imagery products rather than files. Raster products depend on the raster type, and derived products depend on sensor and format specifications, which are listed in the Catalog pane. For example, using Landsat 8 data, you can display and work with pansharpened imagery, which uses 30-meter resolution multispectral bands and the corresponding 15-meter resolution panchromatic band. You can drop the Pansharpen product into the map and use the raster function to dynamically create 15-meter resolution multispectral imagery, add it to the map, and list it in the Contents pane.

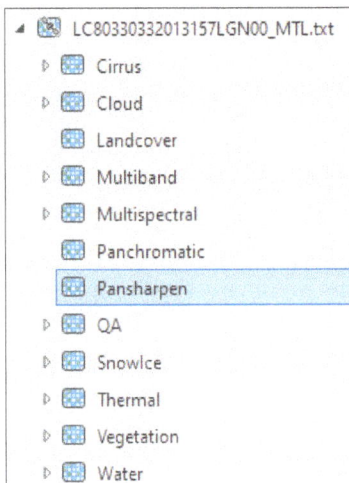

Figure 1-4. Adding a Pansharpen raster product—created on the fly based on metadata—to the map.

Raster functions

Raster functions are operations that apply a dynamic processing operation directly to the pixels of imagery and raster datasets in the display. Only pixels that are visible on your screen are processed as you zoom and pan the image on the fly. Because no intermediate datasets are created, processes can be applied quickly, as opposed to the time it would take to create a processed file on disk.

An advantage of using raster functions is that you can verify and adjust parameter settings and observe the results immediately. The **Raster Functions** pane can be accessed on the **Imagery** tab or the **Analysis** tab. Functions can be applied individually or combined to create function chains. Additional advantages of raster functions include the following:

- **No data duplication:** Because raster functions do not create new datasets, they save storage space and reduce processing time.
- **Real-time processing:** Operations are performed dynamically when the image is viewed or queried, enabling fast analysis and visualization.
- **Efficiency with large datasets:** Raster function templates (RFTs) are useful for processing large image collections (for example, satellite time series) without creating permanent outputs.
- **Flexible and scalable:** Functions can be customized, reused, and shared across projects or users.
- **Creation of raster products:** Raster products are created using raster functions.
- **Integration with image services:** Enables you to interact with processed imagery through web services, supporting cloud-based, enterprise-scale workflows.

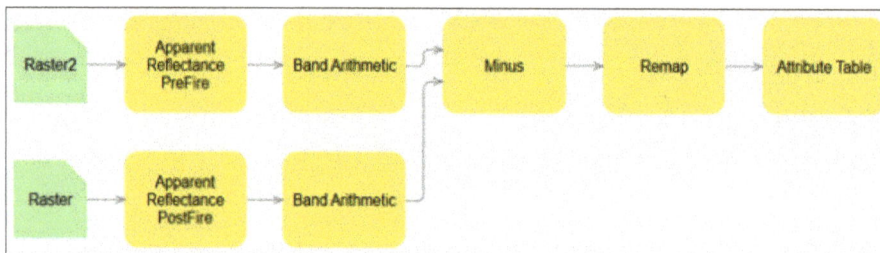

Figure 1-5. Example of a raster function chain (template) built in the Raster Function Editor, and then saved and shared with stakeholders.

Mosaic dataset

Mosaic datasets are used to manage, display, analyze, serve, and share imagery and raster data. The role of a mosaic dataset is to provide the following:

- An image library for management (cataloging, indexing, metadata, searching, and more)
- Dynamic, on-the-fly product generation (mosaicking, processing, and analysis)
- A workflow to shorten the time from sensor to use (quickly ingest, with a dynamic product immediately available)
- Scalability (one to thousands of images)
- Support for homogeneous or heterogeneous collections (one sensor or a mix)
- Dynamic product generation for visualization or analysis

When you create a new mosaic dataset, it is created as an empty container in the geodatabase with some default properties for adding raster data.

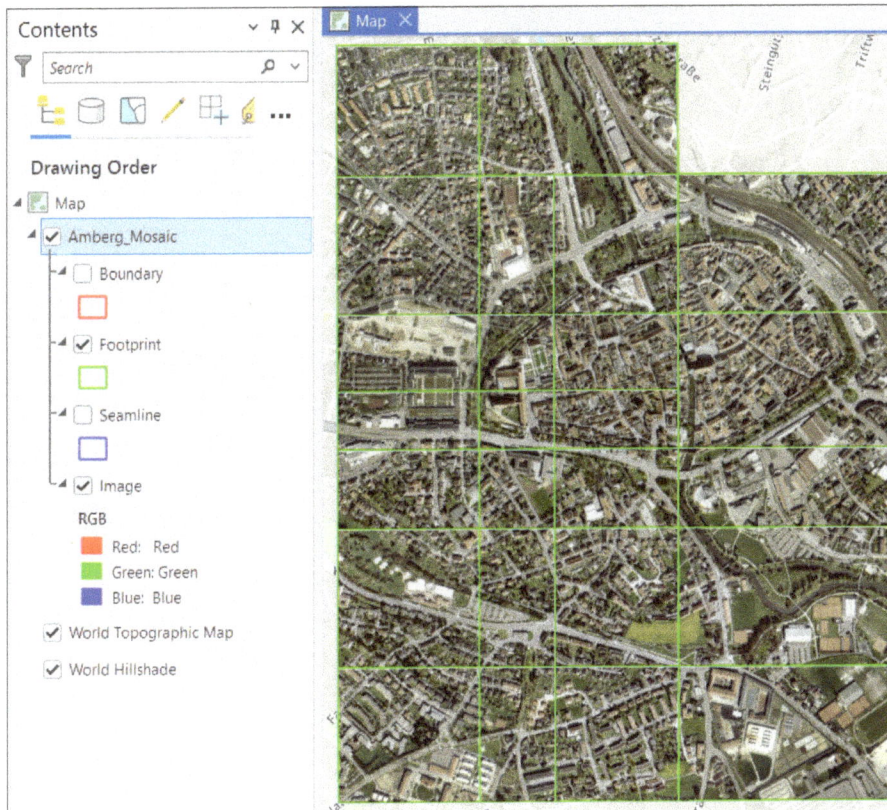

Figure 1-6. A mosaic dataset displayed in the map and listed in the Contents pane.

A mosaic dataset consists of many parts:

- An index that provides the source of the pixels and footprints of the imagery and rasters
- A feature class that defines the boundary
- A set of mosaicking rules that are used to dynamically mosaic the rasters
- A set of properties used to control the mosaicking rules and for image extraction
- A table for logging during data loading and other operations
- A multidimensional table defining variables and dimensions
- A seamline feature class for seamline mosaicking
- Optionally, a stereo table defining stereo models
- Optionally, a color correction table that defines the color mapping for each raster in the mosaic and for the entire mosaic dataset as a single entity

Property	Value
Maximum Size of Requests - Rows	4100
Maximum Size of Requests - Columns	15000
Allowed Compression Methods	None,JPEG,LZ77,LERC
Default Resampling Method	Bilinear Interpolation (for continuous data)
Maximum Number of Rasters Per Mosaic	20
Cell Size Tolerance Factor	0.8
Allowed Mosaic Methods	Seamline,NorthWest,Center,LockRaster,ByAttribute,Nadir,View...
Default Sorting Order	✔
Default Mosaic Operator	First
Blend Width Unit	Pixels
Blend Width	10
Viewpoint Spacing X	600
Viewpoint Spacing Y	300
Always Clip the Raster to its Footprint	✔
Always Clip Overview to its Footprint	☐
Footprint may contain NoData	✔
Always clip the mosaic dataset to its boundary	✔
Apply Color Correction	☐
Minimum Pixel Contribution	1

Image Properties

> Catalog Properties
> Time Properties
> Download Properties

Learn more about mosaic dataset defaults

Figure 1-7. A partial list of properties of a mosaic dataset.

Image services

If your organization has deployed a stand-alone ArcGIS Image Server, you can share several types of image services from ArcGIS Pro. When you publish a service, the server makes it available through a service URL or REST endpoint. Client applications can use the URL to access the service. Services can also be added directly to ArcGIS Pro through an ArcGIS Server connection. See chapters 17, 18, and 19 for details on how to use an image service in ArcGIS Pro.

Data model framework

The following diagram integrates the components of the data model into a comprehensive, coherent framework. It illustrates the relationships between the components of the image data model, from image collection and ingest, image and data management, and processing and sharing within the ArcGIS environment.

Image data models in action

The following scenario illustrates how the image and raster data models enable efficient image management, processing, and sharing in your day-to-day work.

The US Geological Survey (USGS) and other public and private content providers, such as Esri, provide full coverage of the continental United States (CONUS) with Landsat imagery. The collection of Landsat scenes is mosaicked to provide a continuous mosaic dataset stored in online archives. The Landsat mosaic is updated periodically with current imagery.

The data models help make updating the Landsat mosaic straightforward and efficient. New imagery is downloaded from government archives and added to the mosaic using the Add Rasters To Mosaic Dataset geoprocessing tool based on the Landsat data type. The metadata is harvested and used to properly calibrate, project, and add the new imagery to the mosaic dataset. Any raster function templates associated with the Landsat mosaic work with the new data to produce value-added datasets on the fly, such as land cover, vegetation, and urban indices, and raster products, such as multiband and pansharpened imagery. Image overviews, footprints, extent, and other components of the mosaic dataset are automatically updated based on the metadata and mosaic properties. The mosaic dataset updating can be performed in ArcGIS Pro, ArcGIS Online, or ArcGIS Image Server in an ArcGIS Enterprise environment.

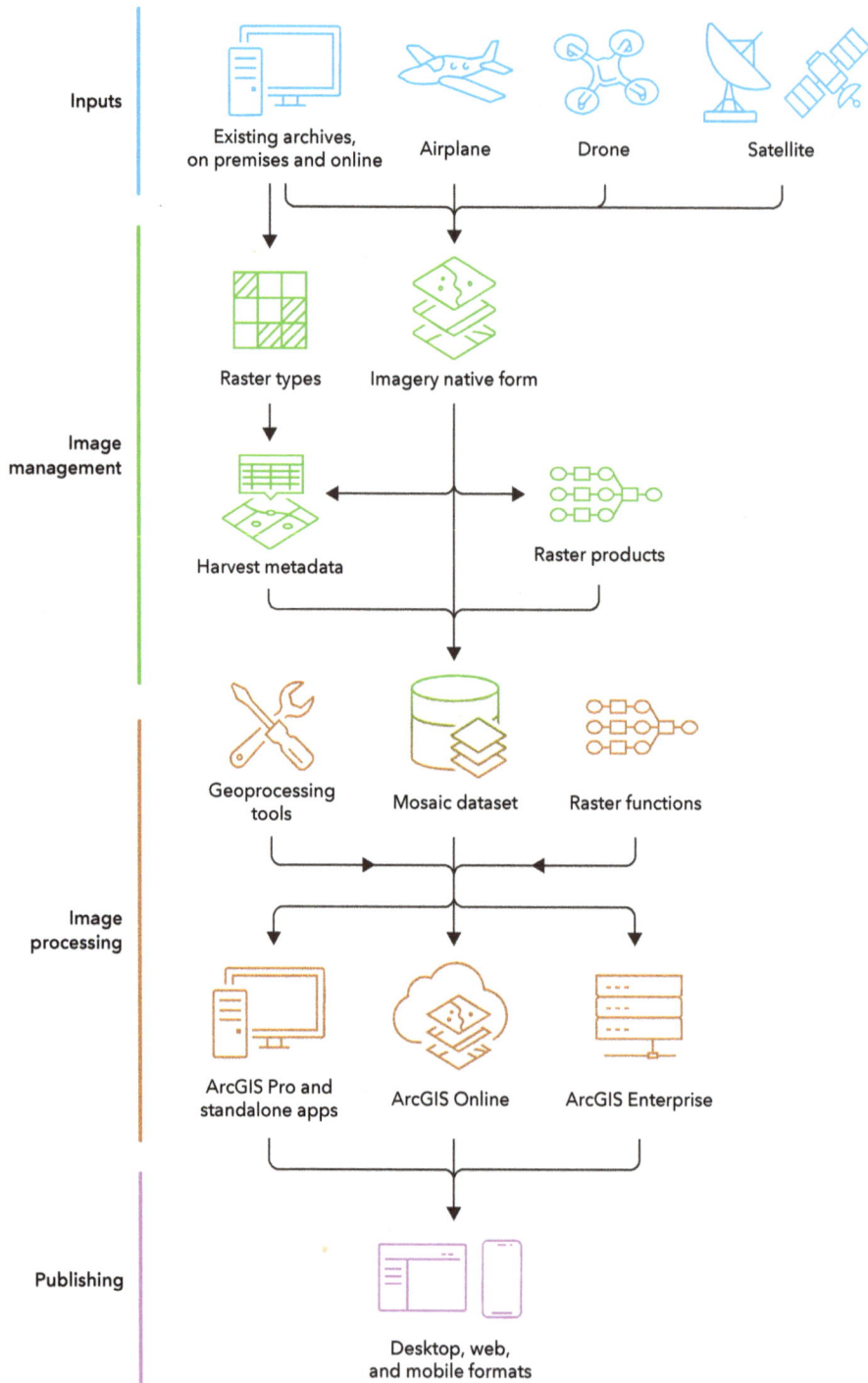

Figure 1-8. The image data model framework.

The mosaic dataset model manages all the Landsat imagery and associated data. New imagery can replace or augment the older imagery because older imagery may be useful for change detection or time series analysis—you do not need to delete it from the mosaic. Using filters and properties, the most current imagery can be displayed and processed, or a different time frame can be specified.

Finally, the Landsat mosaic can be published and shared across your enterprise and used in ArcGIS Pro, web apps, and mobile apps.

Summary

In this chapter, you learned about imagery and remote sensing and how they contribute to a full-functioning GIS. The incorporation of metadata and value-added meta information facilitates image processing that adheres to scientific principles and leads to better efficiency. The seamless integration of image processing and remote sensing with GIS enables deeper geospatial insights and more impactful solutions to complex issues.

The skills presented in this book will equip you with essential capabilities needed to perform common image processing tasks effectively and help you achieve specific outcomes with proficiency to address a variety of scenarios and practical applications you will encounter in your day-to-day work.

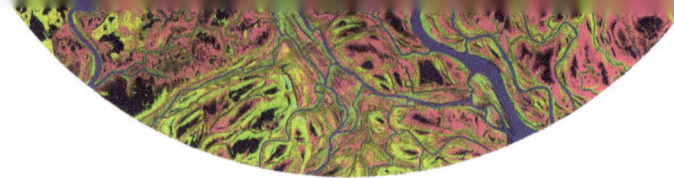

CHAPTER 2
Visualizing imagery

Objectives

- Identify image characteristics using layer properties.
- Change band combinations to aid visual interpretation.
- Apply stretch types to images.
- Compare histogram distributions of an image.

Introduction

Now that you've learned about imagery formats, some imagery terms, and the data model used for imagery and rasters in ArcGIS Pro, let's explore imagery a bit further. In this chapter, you'll learn how to examine the properties, or metadata, of an image and ways to visualize the imagery to identify features. You might think that the metadata included with imagery can't help much with analysis. In fact, it's just the opposite.

Additionally, simple workflows such as modifying an image's band combinations or rendering symbology—such as changing the stretch type of an image—can help you make quick, visual assessments about an area. These observations might otherwise be lost when viewing an image only as a natural color rendering. Sometimes these visual clues can help guide further analysis as you explore an area.

Tutorial 2-1: View a Landsat 8 image in a map view

Raster products are designed so that you can easily display and use imagery. Raster products use the metadata files associated with certain vendor-specific products to optimally render and display images. You'll add a raster product of a Landsat 8 image and explore several of the image properties to learn more about the data.

Open the project and view image data as a raster product

1. Go to links.esri.com/Imagery20Data and download the data for chapter 2.

2. Unzip the folder. On your C: drive, create a folder named Top20Imagery.

> **Note:** Because this is the first chapter with tutorial data, you will need to create the folder **Top20Imagery**. Now and in subsequent chapters, you will download, unzip, and move data for each chapter to this folder.

3. Move the **Top20Imagery_02** folder inside the **Top20Imagery** folder.

    The file path for the folder is now **C:\Top20Imagery\Top20Imagery_02**.

4. Inside the **Top20Imagery_02** folder, double-click **Top20Imagery_Ch02.aprx** to open the ArcGIS Pro project for this chapter.

    The map opens to the region between the Tien Shan mountains and the Taklamakan Desert near the Tarim Basin in northwestern Xinjiang province, China. This area is famous for stunning geological features associated with the Piqiang Fault.

5. On the right, in the **Catalog** pane, expand **Folders** > **Top20Imagery_02**, and then expand the **2018_0930-148_032_L1TP (China)** folder.

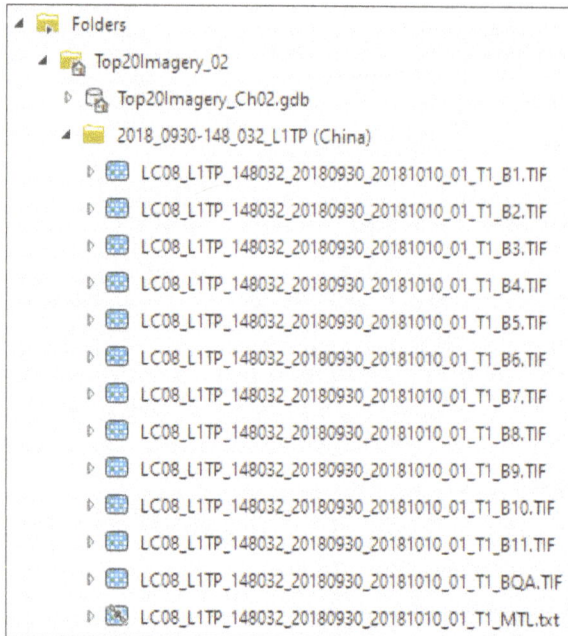

6. Right-click **LC08_L1TP_148032_20180930_20181010_01_T1_MTL.txt** and click **Add To Current Map**.

Tip: A raster product can be quickly identified in the Catalog pane by this icon:

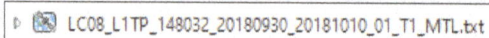

The Landsat 8 image is added to the map and is also visible as a layer in the **Contents** pane. The **Contents** pane gives you information about the layer, such as the image bands that are being rendered in the three color channels. By default, ArcGIS Pro displays Landsat 8 images as natural color. This is why the band combination is shown as red, green, and blue (RGB).

7. On the left, in the **Contents** pane, right-click the **Multispectral_LC08_L1TP_148032_20180930_20181010_01_T1_MTL layer** and click **Properties**.

8. In the **Layer Properties** window, on the **General** tab, click the **Name** field. Delete the current text and type China Landsat 8. Click **Apply**.

    The layer name in the **Contents** pane has updated to reflect your change.

    Note: Often image names used by a vendor contain useful information related to the image, such as the collection date, sensor name, processing information, or even a catalog or purchase order number. However, unless you are familiar with their naming conventions, the default layer names can be difficult to decode. Renaming the layer allows you to quickly and easily see information about the image.

    Next, you'll examine the layer properties to learn more about this image.

Explore the properties of the image

The **Source** tab in the **Layer Properties** window contains a summary of information pulled from the raster product metadata.

1. In the **Layer Properties** window, click the **Source** tab. Click **Raster Metadata** to expand the section.

Expanding this section shows information about the image. This information includes the following fields and values:

Field	Value
Sensor Name	Landsat 8
Product Name	L1TP
Acquisition Date	9/30/2018 05:27:45
Cloud Cover	4.35
Sun Azimuth	155.34
Sun Elevation	44.01

You already knew the sensor name, but not the product type or acquisition date and time. From these fields, you can add more information to the layer name.

2. Using the skills you just learned, add the collection date of Sept 30, 2018, to the layer name.

The name of the layer should be **China Landsat 8 Sept 30, 2018**.

3. Return to the **Source** tab and expand the **Band Metadata** section.

This section shows the multispectral bands collected by the Operational Land Imager (OLI) sensor aboard the Landsat 8 satellite. The bands are named for the portion of the electromagnetic spectrum that they collect.

4. On the right of the window, click the border and expand the window to the right so that all the columns of the table are visible.

In the row for **Red**, you can see information that was collected for the red portion of the electromagnetic spectrum by the Landsat 8 OLI sensor. This information includes the minimum and maximum wavelengths (640 and 670 nanometers, respectively).

5. Collapse the **Band Metadata** section and then expand the **Statistics** section.

This section contains information about the distribution of the pixel values for each band. These include the **Minimum** and **Maximum** values as well as the **Mean** and the **Std. Deviation** of the distribution.

The remaining sections on the **Source** tab (**Raster Information**, **Extent**, **Spatial Reference**) contain information related to the image, such as the spatial extent of the image, the spatial reference system used by the image, and general raster information, such as cell (pixel) size, the number of columns and rows, the image size, radiometric depth and type, whether there are pyramids (reduced resolution datasets) built, and so on.

6. In the **Layer Properties** window, click **OK**.

Tutorial 2-2: Modify the appearance of a Landsat 8 image in a map view

In the previous tutorial, you explored various properties of the Landsat 8 image to get a better understanding of the architecture of the raster product and where to look to find information about your images in ArcGIS Pro. Understanding data distribution can help you with symbolization and rendering. This information, in turn, can help you understand the physical characteristics of features on the ground. In this tutorial, you will modify the appearance of the image in your map using visualization tools and renderings to understand the geographic area of the image.

Change the band combination of an image

One of the quickest ways in ArcGIS Pro to modify the appearance of your image is to change the band combination.

1. In the **Contents** pane, select the **China Landsat 8 Sept 30, 2018** layer.

2. On the ribbon, click the **Raster Layer** tab. In the **Rendering** group, click **Band Combination** and then click **Color Infrared**.

This band combination is frequently used in image analysis to help with vegetation identification. Because the near-infrared band of Landsat 8 is now displayed in the red channel of the three-band composite, the high reflective values resulting from chlorophyl content combined with the low reflective values of the red band displayed in the green channel cause healthy vegetation to display as various hues of red. Other red and pink hues in this image are caused by various geological formations, but the most prominent features are the cultivated agricultural ones at the southern end (*bottom*) of this image. Observe that the band names also changed in the **Contents** pane to help you quickly and easily know what band is being displayed in which color channel.

On your own

Explore the band combinations using the menu. When you finish, return the band combination to the **Color Infrared** band combination.

Change the rendering of an image

There are several other easy ways to modify the appearance of your image. To see these changes more clearly, you will zoom in to an area around the fault line. In addition to the fault line, the various rock types and their alluvial outflows are clearly visible. You can use various rendering and stretch types to enhance these features.

1. Zoom in to the center left of the image.

2. On the **Raster Layer** tab, in the **Rendering** group, click **DRA**.

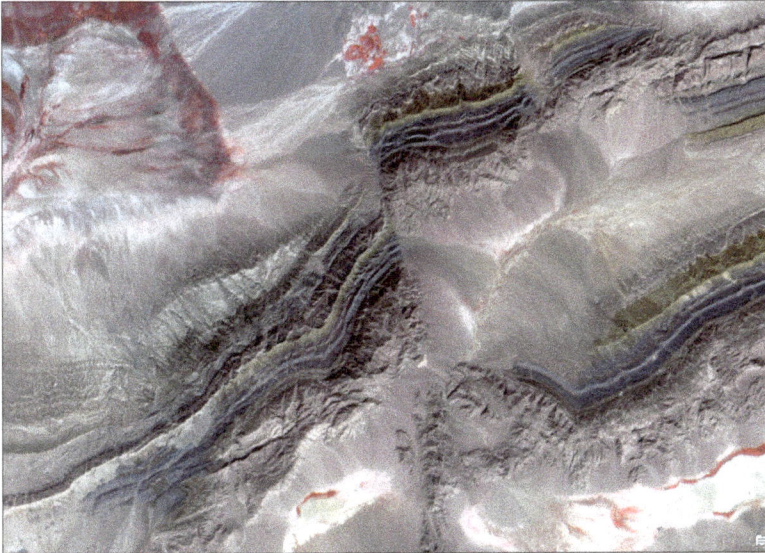

> **Tip:** Some tabs in ArcGIS Pro are contextual, meaning they appear only when necessary. If you can't see the **Raster Layer** tab, in the **Contents** pane, click the **China Landsat 8 Sept 30, 2018** layer.

Clicking **DRA** (Dynamic Range Adjustment) adjusts the rendering to use only the pixel values visible in the map. Often, this results in a dramatically different rendering, sometimes allowing you to discern features that may be washed out or dulled by using the distribution of the pixel values from the entire image.

3. On the **Raster Layer** tab, in the **Rendering** group, click **Stretch type** and then click **Standard Deviation**.

You can also view this stretch based on the histogram values of the pixels visible in the map.

4. On the **Raster Layer** tab, in the **Rendering** group, click **Symbology**.

5. In the **Symbology** pane, on the right of the **Stretch type** field, click the **Histogram** button.

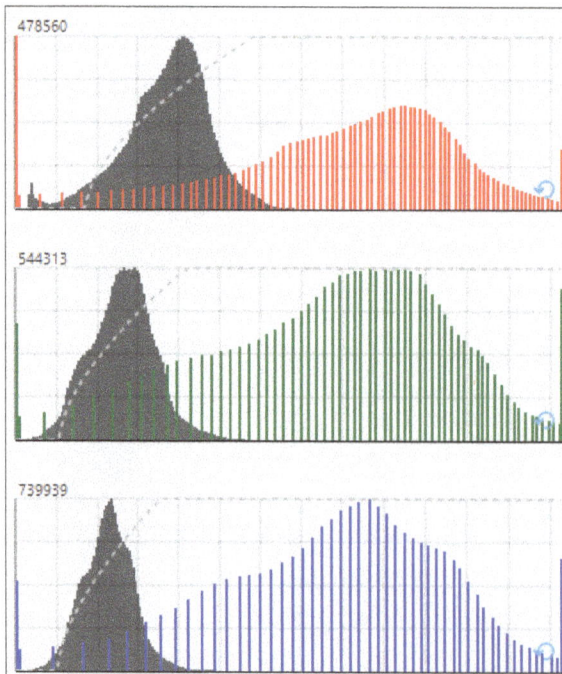

The histograms for each multispectral band are displayed. In this case, the bands are Near-Infrared (red channel), Red (green channel), and Green (blue channel). The gray histograms show the distribution of the real digital number (DN) values of the pixels visible in the map. These values in the gray histogram are the ones you saw in the **Statistics** section of the **Layer Properties** window. The colored histograms show how ArcGIS Pro is stretching the rendered values across the three color channels.

By modifying the stretch type, or even interactively adjusting the breakpoints of the stretch lines, the colored histograms change to reflect the updated rendering distribution.

6. In the **Symbology** pane, for **Stretch type**, click **Histogram Equalize**.

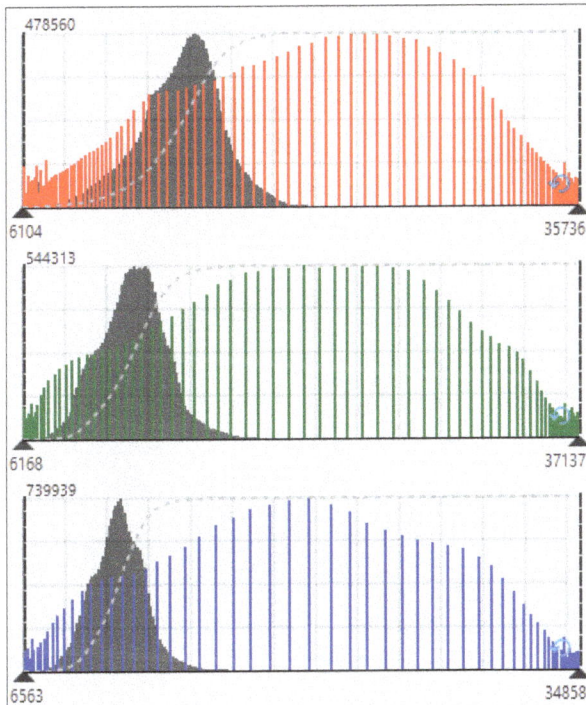

Modifying the stretch type to **Histogram Equalize**—a stretch that redistributes pixel values of an image so that each rendered range contains approximately the same number of pixels—has the practical, and visual, effect of stretching rendered pixel values across the entire range of colors in your map. The difference can be dramatic, but in this image, the result is a much sharper, more richly contrasted image. In this rendering, the various mineral types can be traced much easier through the various alluvial fans.

As you change band combinations or modify rendering types, different details will either emerge or disappear. You can decide, based on your analytic questions or desired output, how best to visualize your imagery.

Take the next step

Change the band combination to one of your choice and modify the rendering of the image by adjusting the histogram and using other skills you just learned. You can also explore adjusting the brightness, contrast, and γ [gamma] settings in the **Enhancement** group on the **Raster Layer** ribbon to see how that changes the visual appearance of your image.

Summary

In this chapter, you learned how to find information about your images that will be important for further analysis. You also explored different renderings of your image by modifying the band combination to perform simple visual analysis. By modifying the rendering of an image, various features or materials otherwise subdued or even hidden can become visible.

Information at your fingertips

Metadata visible in layer properties

Many vendors, both commercial and government, provide extensive metadata with their imagery. Not every vendor provides the same metadata. You are fortunate that the example used in this chapter, Landsat 8, has a lot of valuable metadata associated with it.

Raster metadata

Some metadata may not appear important or relevant for your analysis, but it can be essential. The Sun Azimuth and Sun Elevation, for instance, visible in the Raster Metadata section, are recorded in degrees and are values used in scientific calculations and subsequent applications, such as radiometric calibration.

Often, metadata persists through various processes. For the image used in these tutorials, the cloud cover value of 4.35 is recorded as a percentage. A visual inspection of this image reveals few clouds, certainly no more than 4%. Recall, however, that this image is only a small portion of a larger, full image collection. This derived value provided by USGS is based on the metadata of the entire image. When the image was processed to include only this area, a real-time estimate of the new cloud cover wasn't derived.

Raster information

Reviewing the Raster Information section for this Landsat 8 image reveals that the image is unsigned short (Pixel Type) 16 bit (Pixel Depth). The Landsat 8 OLI sensor collects spectral information across a 12-bit dynamic range (radiometric resolution) resulting in 2^{12} or 4,096 possible data values. However, these collected data values are rescaled to a 16-bit data value range by the vendor (USGS) before product delivery for use in remote sensing or GIS apps. This results in a pixel having 2^{16} possible data values (65,536). These resulting values are called DN values. This is why, when you observe the maximum values in the Statistics section, all are greater than 4,096.

Band metadata

The Band Metadata section contains information used to help understand the spectral nature of your image. Often, it will also contain information useful in calibrating the image radiometrically. The Landsat 8 image you used in this chapter contains information for each band associated with Gain values for reflectance and radiance measures as well as Bias values for both reflectance and radiance values. The Radiance values are in watts per square meters per micron. This unit is important for use in atmospheric calibration calculations. Each band section will have slightly different values based on collection differences and sensitivities.

Rendering options

DRA rendering for dynamic range adjustment is based on a particular stretch type being applied based on your selection. By default, ArcGIS Pro uses a stretch type of Percent Clip. After you modify the stretch type, as you did in tutorial 2-2, the rendered image automatically applies the DRA to this new stretch type. To render the image using the full range of pixel values of an image, you can turn off DRA and perform your visual assessment and analysis that way.

Ready-made band combinations, such as the ones you used in the tutorials, are available when a vendor-supplied raster product contains the appropriate band metadata, such as wavelength range. Different band combinations are visible depending on what bands are available for a particular image or raster dataset. In other words, a four-band image will not have all the options you saw in this chapter, only two: Natural Color and Color Infrared. Additionally, different images with spectral ranges different from Landsat 8 may have different options available as well.

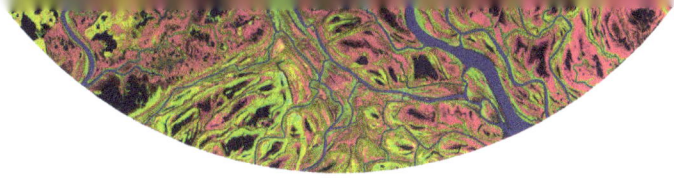

CHAPTER 3
Working with raster functions

Objectives

- Apply raster functions to imagery.
- Build a raster function template for image analysis.

Introduction

In this chapter, you'll learn how to use raster functions for analysis. Raster functions are operations that process the pixels of an image or raster dataset directly in your map, using on-the-fly processing and calculations. These calculations are applied to the pixels of the image as displayed, so only the pixels visible on your screen are processed. Raster functions are not permanent and do not affect the original imagery. Raster functions perform the specified task and create a new raster layer in the project. This new layer output is displayed in the map based on the parameters specified in the raster function.

Tutorial 3-1: Apply a raster function to an image

In this tutorial, you'll learn how to apply a raster function to your imagery. There are several ways to apply raster functions in ArcGIS Pro. Some common processes are available directly on the **Imagery** tab as indexes. Indexes are mathematical operations performed using spectral bands to emphasize or enhance specific phenomena in an image. The advantage of using indexes, sometimes called band ratios, is that the spectral characteristics of an object, such as healthy or unhealthy vegetation, are emphasized rather than the intensity or brightness of an object, regardless of the illumination conditions of an image. They are tied to specific bands of the electromagnetic spectrum. One of the most common ways that indexes are used is for comparison of the same object across multiple images over time.

You can also access the full range of raster functions by using the **Raster Functions** pane. Raster functions can be used for analysis, visualization, radiometric calibration, geometric correction, and even data management.

Calculate an index using a raster function

Brush fires are not a new phenomenon in Australia. Indeed, the geography of Australia is conducive to large brush fires. Rangers at Karlamilyi National Park in Western Australia must continually be on guard and understand the fire danger. Using a pair of Landsat 8 images collected a year apart showing fire activity northwest of the park, you will use this imagery to understand and track these fires.

Download the tutorial data and set up the project

1. Go to links.esri.com/Imagery20Data and download the data for chapter 3.

2. Unzip the folder to **C:\Top20Imagery**.

> **Note:** In the second chapter, you created a folder named **Top20Imagery** on your C: drive. If you haven't done that, create that folder now. Now and in subsequent chapters, you will download and unzip data for each chapter to this folder.

3. Inside the **Top20Imagery_03** folder, double-click **Top20Imagery_03.aprx** to open the ArcGIS Pro project for this chapter.

View the imagery

To begin your assessment of wildfire areas in Karlamilyi National Park, you will add the 2019 Landsat 8 image first.

1. In the **Catalog** pane, expand **Folders > Top20Imagery_03.** Expand the **2019_0202-111_075_L1TP (Australia)** folder.

2. Right-click the **LC08_L1TP_111075_20190202_20190206_01_T1_MTL.txt** raster product and click **Add To Current Map**.

The image appears on the map. You will use a predefined raster function to identify areas of wildfire activity.

3. In the **Contents** pane, confirm the **Multispectral_LC08_L1TP_111075_20190202_20190206_01_T1_MTL** layer is selected.

4. On the **Imagery** tab, in the **Tools** group, click **Indices**. Under **Landscape**, click **NBR**.

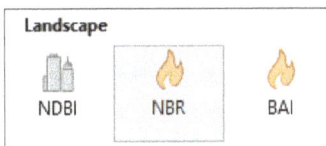

5. In the **NBR** dialog box, under **Near Infrared Band Index**, ensure that
    5 – NearInfrared (850-88nm) is selected. For **Shortwave Infrared Band
    Index**, click **7 – ShortwaveInfrared_2 (2110-2290 nm)**.

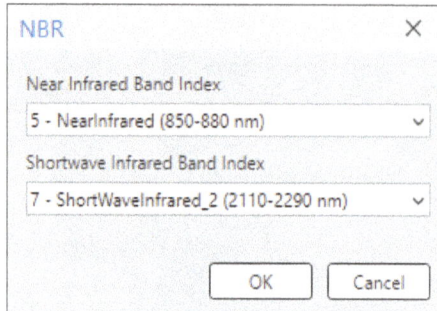

A raster function is performing this calculation, which is why it can derive this
information so quickly.

6. Click **OK**.

The resulting layer is the result of a raster function of the Normalized Burn Ratio (NBR). This index uses a predefined formula (NIR − SWIR) / (NIR + SWIR) to calculate this result on the fly. The dark areas indicate locations that have burned.

Apply a new raster function

1. In the **Catalog** pane, expand the **2020_0308-111_075_L1TP (Australia)** folder.

2. Right-click the **LC08_L1TP_111075_20200308_20200822_02_T1_MTL.txt** raster product and click **Add To Current Map**.

3. In the **Contents** pane, confirm that the **Multispectral_LC08_L1TP_111075_20200308_20200822_02_T1_MTL** layer is selected.

4. On the **Raster Layer** tab, in the **Rendering** group, click **Band Combination** and then click **Vegetation Analysis**.

This band combination is referred to in ArcGIS Pro as the Vegetation Analysis band combination. Because it uses the Near-Infrared (NIR) and Shortwave Infrared (SWIR) bands, it is also useful for identifying brushfire, or wildfire, scars. A rule of thumb, when observing these features in this band combination, is that the darker and more reddish-brown a feature is, the more recent the burn. As burn scars age, they become more red or pink. As they become revegetated, they will start to appear greener.

You'll now apply the same index to the 2020 image using a **Band Arithmetic** raster function.

5. On the **Imagery** tab, in the **Analysis** group, click **Raster Functions** to open the **Raster Functions** pane.

   Unlike other common indexes such as the Normalized Difference Vegetation Index (NDVI), the NBR index isn't predefined here as a separate function in the **Raster Functions** pane. It is, however, available as part of the **Band Arithmetic** function.

6. In the **Raster Functions** pane, expand the **Math** section and click **Band Arithmetic**.

   The **Band Arithmetic** function allows you to perform mathematical operations using different bands of your imagery. Often, if you can't find a ratio or index in the **Raster Functions** pane, there's a good chance it's available in the **Band Arithmetic** section. You can even build your formulas and include them as **User Defined** indexes. To assess burn severity, the NBR formula is already available as a predefined method for calculation.

7. In the **Band Arithmetic Properties** function, for **Raster**, click the **Multispectral_LC08_L1TP_111075_20200308_20200822_02_T1_MTL** layer.

8. For **Method**, click **NBR**.

   Tip: You may need to scroll up through the list to find NBR.

   As before, the bands you will use to calculate the NBR are the NIR and SWIR bands. For Landsat 8, this corresponds to bands 5 (NIR) and 7 (SWIR2). Because ArcGIS Pro has this formula predefined, you'll only need to designate which bands you want to use.

9. For **Band Indexes**, type 5 7 with a space in between.

> *Important:* *The input for the bands is a space-delimited list. In other words, when adding the band numbers to any index, make sure there is a space between the identified bands and not a comma.*

Raster

| Multispectral_LC08_L1TP_111075_20200308_20200822_02_ ˘ ✕ | 📁 |

Method

| NBR | ✕ |

Band Indexes

| 5 7 |

Input: NIR SWIR
Output: (NIR - SWIR) / (NIR + SWIR)

10. At the bottom of the pane, click **Create new layer**.

The new layer is automatically added to your map. You can also see the calculated range of values in the **Contents** pane.

11. On the **Raster Layer** tab, in the **Compare** group, click **Swipe**.

12. Using the **Swipe** tool, visually examine the **Band Arithmetic** and **NBR** layers. Adjust the visibility of the other layers in the **Contents** pane as needed.

13. After exploring and comparing these two raster function results, in the **Contents** pane, right-click each raster function layer and click **Remove**.

Tutorial 3-2: Build a raster function chain

You can combine several raster functions together into a raster function chain to perform more thorough analysis. These chains can then be saved as a raster function template and applied to other raster types, such as mosaic datasets or image services. If your image analysis workflow contains repeatable steps, raster function templates provide an easy way to build and share your workflow with others to be used on different datasets with similar image properties.

In this tutorial, you will build a raster function chain to analyze the burn severity in Karlamilyi National Park. In addition, you will use the result of this comparison to identify the extent of the burn areas as well as areas of potential vegetation regrowth in the previous years' burn extents.

Add layers and functions to a raster function template

1. On the **Imagery** tab, in the **Analysis** group, click **Function Editor**.

2. From the **Contents** pane, drag the **Multispectral_LC08_L1TP_111075_ 20190202_20190206_01_T1_MTL** and the **Multispectral_LC08_L1TP_111075_ 20200308_20200822_02_T1_MTL** layers into the **Function Editor**.

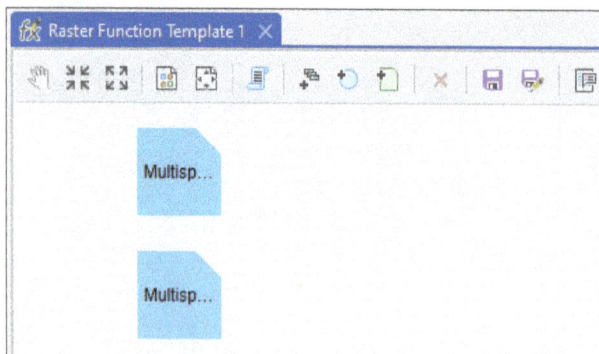

Now that you've added the two images, you will begin to add raster functions to your chain.

3. From the **Raster Functions** pane, drag the **Band Arithmetic** function into the **Function Editor**, to the right of the first raster layer.

4. Drag a second **Band Arithmetic** function into the **Function Editor**, to the right of the second raster layer.

5. In the **Function Editor**, drag a line from the **Multispectral_LC08_L1TP_111075_20190202_20190206_01_T1_MTL** raster to one of the **Band Arithmetic** functions and connect as **Raster**.

6. Repeat this process for the **Multispectral_LC08_L1TP_111075_20200308_20200822_02_T1_MTL** raster.

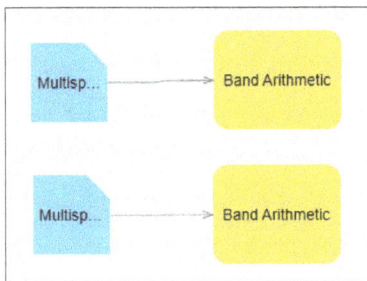

7. In the **Raster Functions** pane, from the **Math** section, drag the **Minus** function into the **Function Editor**.

8. In the **Raster Functions** pane, in the search bar at the top, type Remap.

9. Drag the **Remap** function into the **Function Editor**.

10. In the **Raster Functions** pane, search for Attribute Table and drag the **Attribute Table** function into the **Function Editor**.

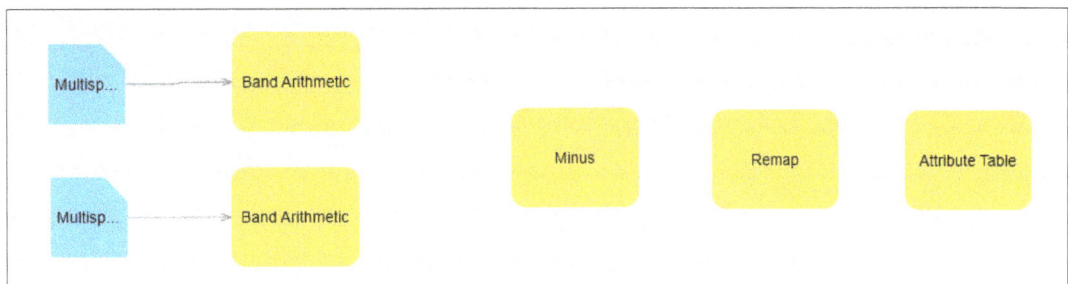

Now that you have your raster layers and the raster functions of your workflow, you will set up the appropriate parameters for your analysis.

Modify the functions in a raster function template

1. In the **Function Editor**, double-click the first **Band Arithmetic** function.

    The **Band Arithmetic Properties** window allows you to set specific parameters for the raster function. Instead of using the NBR function as you did earlier, you'll use a different formula. This formula is called the Normalized Burn Ratio 2 (NBR2). This ratio exploits the sensitivity to water in the shortwave infrared portion of the electromagnetic spectrum and is useful to help determine moisture levels in any vegetation regrowth in burn scars. Analysis of this type helps scientists understand how well a region may be recovering following brush fires or wildfires.

2. Set **Method** to **User Defined**.

    Using a user-defined formula allows you to configure the formula for the NBR2. The NBR2 formula is similar to the NBR formula but substitutes the SWIR1 band in place of the NIR band. This new formula is (SWIR1 – SWIR2) / (SWIR1 + SWIR2). You'll use band 6 and band 7 for your new index.

3. In the **Band Indexes** field, type (B6–B7)/(B6+B7).

Raster
<Stretch Function.OutputRaster>
Method
User Defined
Band Indexes
(B6-B7)/(B6+B7)

4. In the **Band Arithmetic Properties** window, click the **General** tab.

5. For **Name**, type Pre-Fire Image (NBR2).

Name
Pre-Fire Image (NBR2)
Description
Calculates indexes using predefined formulas or a user-defined expression.

6. Click **OK**.

7. Repeat this process for the second **Band Arithmetic** function. However, name this one Post-Fire Image (NBR2).

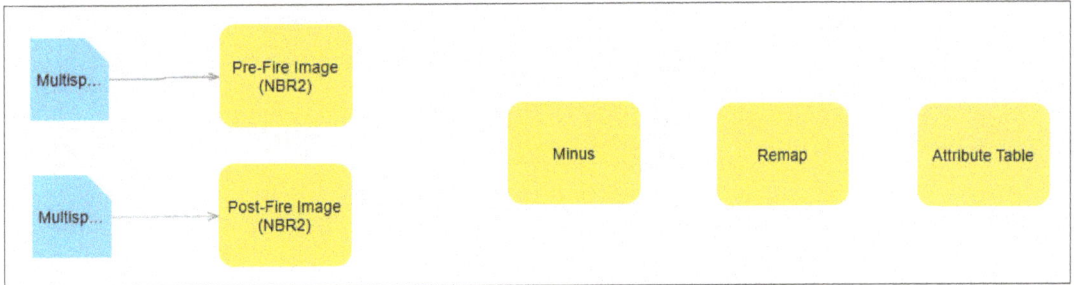

8. In the **Function Editor**, drag a line from the **Pre-Fire Image (NBR2)** function by hovering on the right of the Pre-Fire input to activate the endpoint, and then press and hold as you connect to the **Minus** function as **Raster**.

9. Drag a line from the **Post-Fire Image (NBR2)** function and connect to the **Minus** function as **Raster2**.

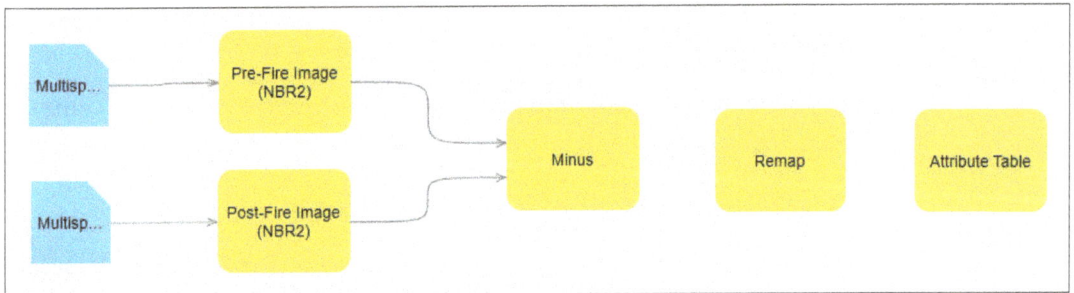

The **Minus** function will subtract the NBR2 index result from 2020 from the NBR2 index result from 2019. This will give you the relative difference (change) between the two indexes. Through examining the results, you can identify threshold values to identify potential categories of change. You can use these values to remap the continuous data values from the index result into discrete categories. You'll set these values using the **Remap** function and then assign the categories using the **Attribute Table** function.

10. Drag a line from the **Minus** function and connect to the **Remap** function as **Raster**.

11. In the **Function Editor**, double-click the **Remap** function.

12. In the **Remap Properties** window, ensure that the **Remap Definition Type** is set to **List** and update the **Minimum**, **Maximum**, and **Output values** using the following table.

Minimum	Maximum	Output
-0.322	-0.02	1
-0.02	0.007	2
0.007	0.014	3
0.014	0.161	4

Note: These values are for educational purposes only.

13. Click the **General** tab and set the **Output Pixel Type** to **8 Bit Unsigned.**

The **Attribute Table** function, which will be the next function in the chain, can only take input rasters that are 8 bit. This is why you set the **Output Pixel Type** to **8 Bit Unsigned.**

14. Click **OK**.

15. Drag a line from the **Remap** function and connect to the **Attribute Table** function as **Raster**.

16. In the **Function Editor**, double-click the **Attribute Table** function.

17. In the **Attribute Table Properties** window, set the **Table Type** to **Manual** and update the **Value** and **Class Name** values and the **Color** patch using the following table:

Value	Class Name	Color
1	2019 Fire Scar Recovery	Leaf Green (Column 7, Row 5)
2	Minimal Change 1	Macaw Green (Column 6, Row 4)
3	Minimal Change 2	Cantaloupe (Column 3, Row 2)
4	Recent Fire Activity	Mars Red (Column 2, Row 3)

Click **OK**.

18. In the **Function Editor** toolbar, click the **Save As** button.

19. In the **Save As** window, apply the following settings:
    - **Name**: NBR2 Severity Comparison
    - **Category**: Custom
    - **Sub-Category**: Custom1
    - **Description**: A raster function chain to calculate burn severity and potential regrowth areas using the Normalized Burn Ratio 2 formula with pre- and postimagery.

20. Click **OK**.

Run the new raster function template

You can now use your new raster function template to identify areas of burns and potential regrowth areas between these two images.

1. In the **Raster Functions** pane, clear the search field. Click the **Custom** tab and then click the **NBR2 Severity Comparison** function template.

   You built your function template to use the two Landsat 8 images from the **Contents** pane. Because of this, you do not need to set any input files or any other settings. You can run the entire function chain from here.

2. In the **Raster Functions** pane, click **Create new layer**.

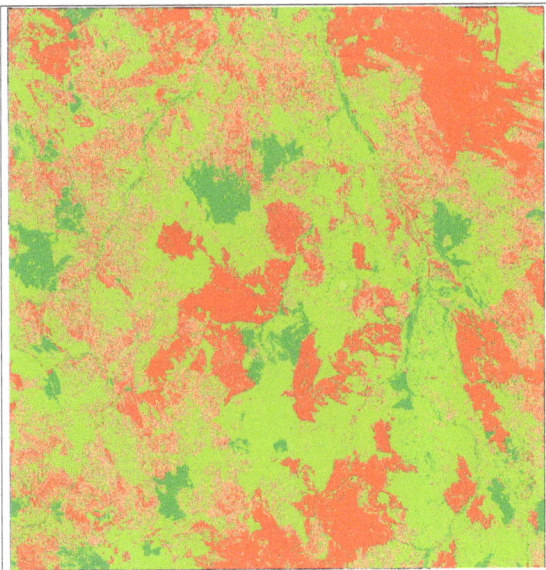

You can modify the raster function template to select different images.

3. On the **Function Editor** toolbar, click **Add Raster Variable** and drag a line to the **Pre-Fire Image (NBR2)** function.

4. Double-click the **Pre-Fire Image** function, and in the **Band Arithmetic Properties** window, click the **Variables** tab. Under **Raster**, check the box for **IsPublic**. Click **OK**.

   Doing this will automatically disconnect the preset raster and let you select which raster to process.

5. Repeat this process for the **Post-Fire Image (NBR2)** function.

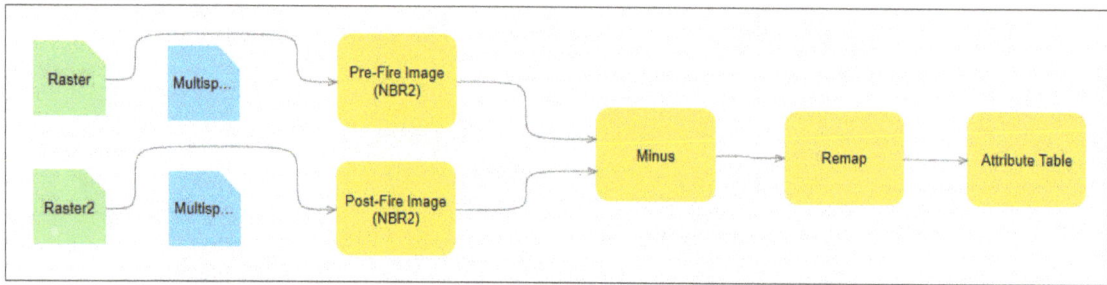

6. In the **Function Editor**, click **Save**.

7. In the **Raster Functions** pane, on the **Custom** tab, click the **NBR2 Severity Comparison** function template.

   Now when you run this raster function template, you can select images of your choice. These can be images visible in the **Contents** pane and the map or from another folder location.

   You can clean up your raster function template by removing the existing Landsat 8 layers.

8. In the **Function Editor**, right-click the **Multispectral_LC08_L1TP_111075_20190202_20190206_01_T1_MTL** raster and click **Delete**.

9. Repeat for the **Multispectral_LC08_L1TP_111075_20200308_20200822_02_T1_MTL** raster.

10. On the **Function Editor** toolbar, click the **Auto Layout** button.

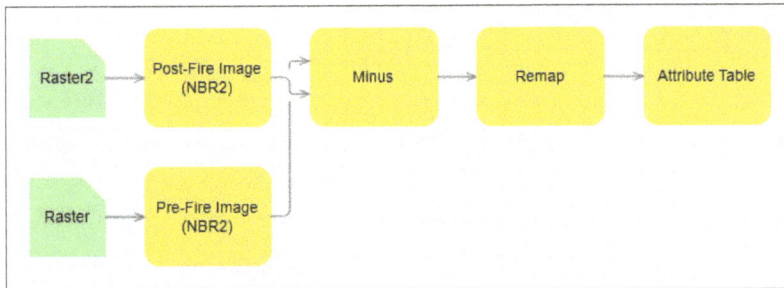

11. On the **Function Editor** toolbar, click **Save**.

12. On the **Quick Access** toolbar, click **Save Project**.

> **Resource tip:** A copy of the raster function template used in this tutorial is available in the data for this chapter named **NBR-2 Raster Function Template.rft.xml**.

Take the next step

You can combine these results with other raster functions for vegetation analysis, such as the Soil-Adjusted Vegetation Index (SAVI) or the Modified Soil-Adjusted Vegetation Index (MSAVI) to measure vegetation activity in the potential new growth areas. You could apply various moisture indexes to mask out areas of water or riparian vegetation, which are causing confusion in this initial result. There are also other indexes available in the Band Arithmetic raster function that might prove useful to an analysis of this type.

Summary

In this chapter, you learned how to apply raster functions to imagery and how to chain multiple raster functions together to build a raster function template for more thorough analysis.

Information at your fingertips

The Normalized Burn Ratio, or NBR, works by exploiting the relative difference between spectral responses of material on the earth's surface. Burn areas have a high spectral response in the shortwave infrared portion of the electromagnetic spectrum (EMS), while at the same time having a low spectral response in the near infrared. Conversely, healthy vegetation has a high return in the near infrared and a correspondingly low spectral response in the shortwave infrared. The SWIR bands benefit from some cloud and smoke penetration. The spectral ranges of both the SWIR_1 band (1570.0 nm – 1650.0 nm) and SWIR_2 band (2110.0 nm – 2290.0 nm), as well as the NIR band (850.0 nm – 880.0 nm) are sensitive to water and moisture. Water absorbs most of the energy in these portions of the spectrum. Understanding these and other spectral characteristics allows you to understand how indexes work.

It is recommended practice when calculating indexes to calibrate your images to reflectance. In ArcGIS Pro, you can use the Apparent Reflectance raster function to ensure that your image uses metadata (built-in sensor information) to calibrate differences in brightness resulting from varying illumination or terrain factors. This raster function calibrates your image to a top of atmosphere relative reflectance. In this lesson, you used Level 1, 16 Bit Unsigned DN values. You will learn more about radiometric calibration and how to calibrate images to reflectance values in chapter 5.

For information on the Normalized Burn Ratio 2, see links.esri.com/LandsatNBR2.

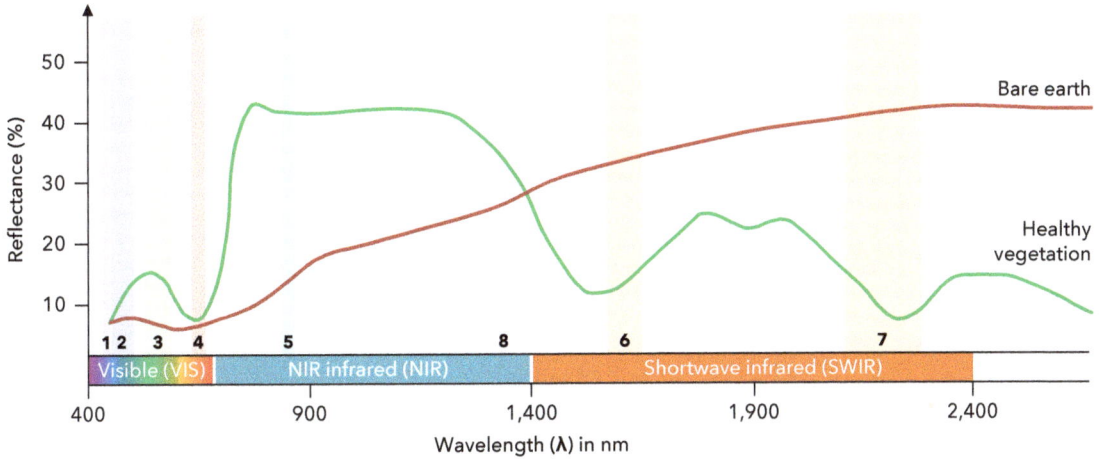

A spectral signature of healthy vegetation has a unique shape. The NIR portion of the EMS is very sensitive to chlorophyll content and plant cell structure causing a high spectral response. By comparison, vegetation has a lower response in the SWIR portion of the spectrum. Landsat 8 collects and records these spectral responses in Band 5 (NIR) and Band 7 (SWIR_2).

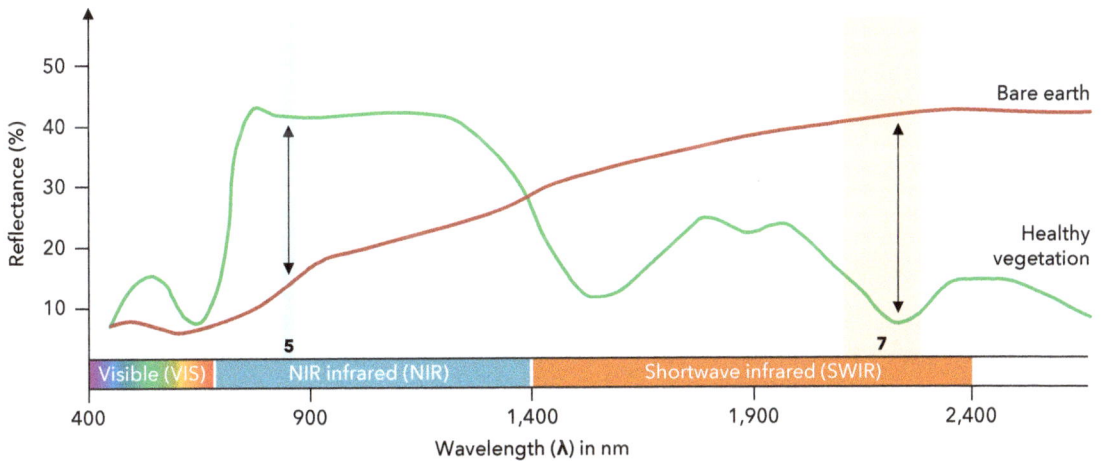

Burned areas spectral signatures also have a unique shape. Because of the lack of healthy vegetation in burned-out areas, there is a much lower response in the NIR portion of the spectrum. Conversely, because of chemical changes in the soil, charcoal concentrations in the burn scar, and other factors, burned areas reflect very highly in the SWIR portion of the EMS. In this graphic, a burn scar spectral response curve is overlaid with the healthy vegetation spectral response curve.

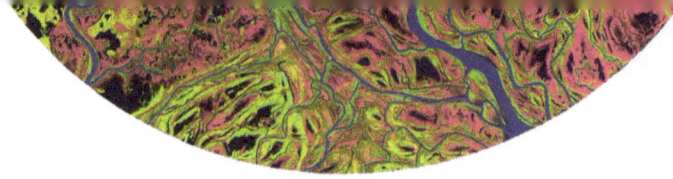

CHAPTER 4
Creating a mosaic dataset

Objectives

- Build a mosaic dataset.
- Create a mosaic dataset container.
- Add imagery to the mosaic dataset.
- Create footprints, build overviews, and explore functionality in ArcGIS Pro.
- Explore attributes of the mosaic dataset and individual items.

Introduction

In this chapter, you'll learn how to create a mosaic dataset comprising two Landsat 8 images. Mosaic datasets are fundamental to practical image management, visualization, and advanced analysis.

A mosaic dataset is a data structure used to manage, display, analyze, and share collections of image data, such as satellite, aerial, and drone images. When you create a mosaic dataset, it is an empty container in the geodatabase with some default properties to which you can add image data. A mosaic dataset container allows collections of imagery to be as follows:

- Seamlessly mosaicked together for visualization

- Stored as references to the actual image files stored on disk
- Queried and analyzed as a single, unified layer

Mosaic datasets support the following:

- Different file types and resolutions
- Dynamic mosaicking rules, such as sort by date and view closest to center
- On-the-fly processing, including pansharpening, band math, and orthorectification

Tutorial 4-1: Create the mosaic dataset container

You will create a mosaic dataset comprising two Landsat 8 Operational Land Imager (OLI) satellite scenes. After the imagery is added, you will define image footprints and create overviews by removing NoData pixels from the scenes.

Download the tutorial data and set up the project

1. Go to links.esri.com/Imagery20Data and download the data for chapter 4.

2. Unzip the folder to **C:\Top20Imagery**.

   Note: In the second chapter, you created a folder named **Top20Imagery** on your C: drive. If you haven't done that, create that folder now. Now and in subsequent chapters, you will download and unzip the data for each chapter to this folder.

3. Inside the **Top20Imagery_04** folder, double-click **Top20Imagery_04.aprx** to open the ArcGIS Pro project for this chapter.

   The Landsat 8 dataset comprises two directories containing the scenes, named **LC08_L2SP_034032_20220926_20221004_02_T1** and **LC08_L2SP_033032_20220903_20220913_02_T1**. These directories contain each band of the Landsat 8 imagery and the associated metadata.

4. In the **Catalog** pane, expand **Folders > Top20Imagery_04 > LC08_L2SP_034032_20220926_20221004_02_T1**.

   The image and metadata files are listed.

5. Expand **LC08_L2SP_034032_20220926_20221004_02_T1_MTL.txt**. Right-click the **Multiband** raster product and click **Add To Current Map**.

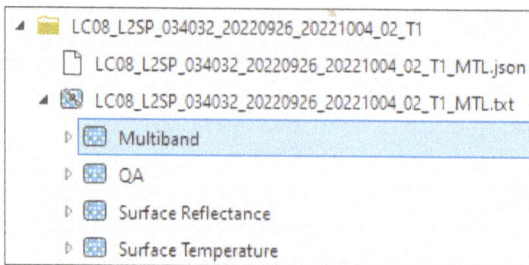

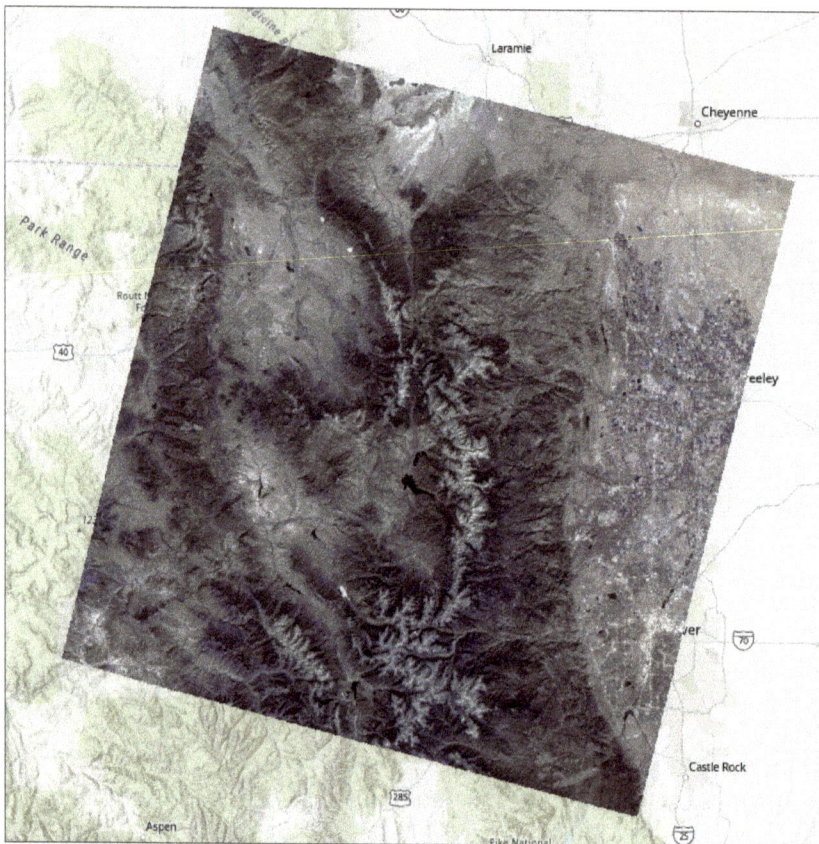

The image is added to the map and listed in the **Contents** pane.

6. Zoom and pan to explore the Landsat imagery.

7. On the ribbon, click the **Raster Layer** tab and explore the **DRA**, **Band Combination**, and **Stretch type** options.

8. In the **Contents** pane, uncheck the box next to the layer name to turn off visibility.

Create the mosaic dataset container

1. In the **Catalog** pane, under **Folders > Top20Imagery_04**, right-click the **Top20Imagery_04** geodatabase. Hover over **New** and click **Mosaic Dataset**.

   The **Create Mosaic Dataset** tool appears.

2. In the **Create Mosaic Dataset** tool, apply the following settings:
   - **Mosaic Dataset Name**: COfrontrangeMosaic
   - **Coordinate System**: Multiband_LC08_L2SP_034032_20220926_20221004_02_T1_MTL

     *Important: When you select the layer in the drop-down list, the coordinate system of the Landsat scene is automatically populated as **WGS_1984_UTM_Zone_13N**.*

   - **Product Definition**: Landsat OLI
   - **Pixel Properties**: 16-bit unsigned

   > Output Location
   > `Top20Imagery_04.gdb`
   >
   > Mosaic Dataset Name
   > `COfrontrangeMosaic`
   >
   > Coordinate System
   > `WGS_1984_UTM_Zone_13N`
   >
   > Product Definition
   > `Landsat OLI`
   >
   > \> Product Properties
   > ⌄ Pixel Properties
   > Pixel Type
   > `16-bit unsigned`

3. Click **Run**.

   The empty **COfrontrangeMosaic** mosaic dataset container is created once you run the tool and appears in the **Contents** pane.

4. In the **Contents** pane, click the **COfrontrangeMosaic** layer.

   On the ribbon, two new tabs are displayed, **Mosaic Layer** and **Data**. Next, you will add your imagery to the mosaic dataset.

Add imagery to the mosaic dataset

You will specify two Landsat 8 scenes to be added to the mosaic dataset.

1. In the **Catalog** pane, in the **Top20Imagery_04.gdb** folder, right-click the empty **COfrontrangeMosaic** mosaic dataset you just created and click **Add Rasters**.

   The **Add Rasters To Mosaic Dataset** tool appears with a few settings already populated, such as the **Mosaic Dataset** name, the **Raster Type**, and the **Processing Template** settings. You will change some of the settings.

2. For **Processing Templates**, click **Multiband**.

   The **Raster Type** setting is automatically populated with the **Multiband** raster type.

   > Note: ArcGIS Pro supports the most common sensors. When you select the type of imagery you are working with from the list, ArcGIS uses information in the metadata of the imagery to intelligently manage it. Selecting a product definition automatically populates the fields for the specifications of each band.

3. Under **Input Data**, confirm **File** is selected. Click the **Browse** button to open the **Input Data** dialog box.

   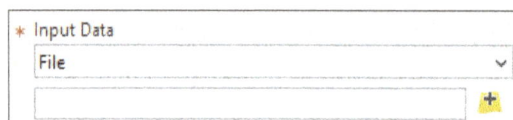

4. Double-click **Folders > Top20Imagery_04 > LC08_L2SP_034032_20220926_20221004_02_T1**. Select the Landsat 8 file **LC08_L2SP_034032_20220926_20221004_02_T1_MTL.txt** and click **OK**.

   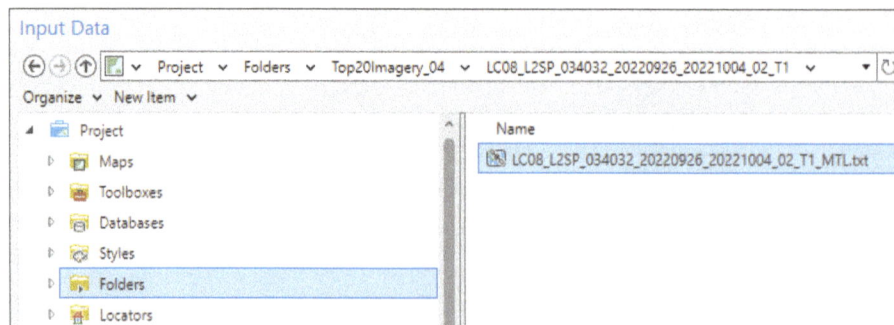

The file is displayed in the **Input Data** list.

5. Open the **Input Data** dialog box again. Navigate to the **LC08_L2SP_033032_20220903_20220913_02_T1** folder, select the Landsat 8 file **LC08_L2SP_033032_20220903_20220913_02_T1_MTL.txt**, and add it.

    The files are displayed in the list of **Input Data**.

6. In the tool, click **Raster Processing** to expand it and check the **Calculate Statistics** box.

7. Click **Run**.

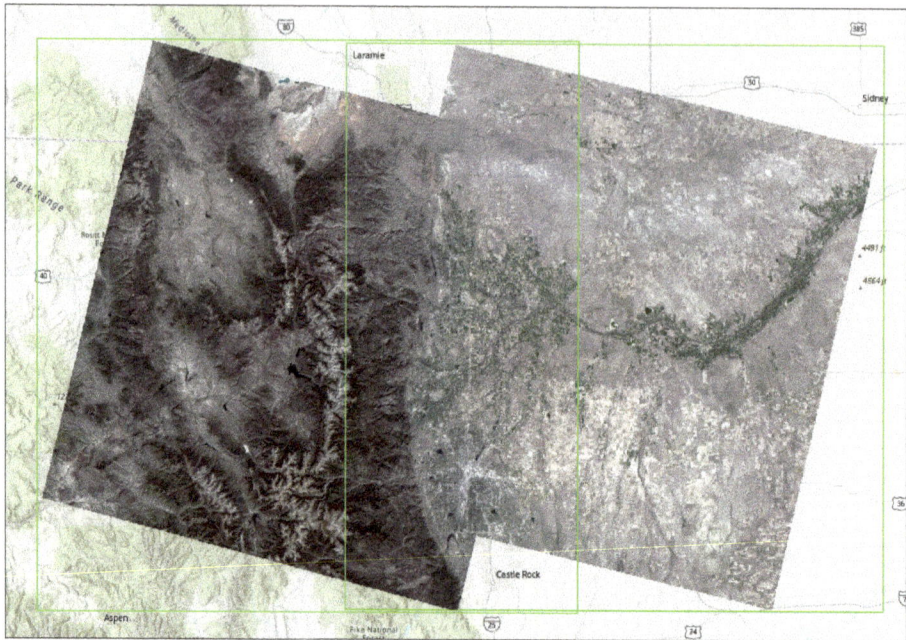

The **COfrontrangeMosaic** mosaic dataset is updated to contain the Landsat 8 imagery and associated metadata.

8. On the map, pan and zoom in the mosaic dataset.

When you zoom far out, the image overviews are not displayed. These overviews will be created in a later step.

Build footprints and overviews

1. In the **Catalog** pane, right-click the mosaic dataset, hover over **Modify**, and click **Build Footprints**.

2. In the **Build Footprints** tool, make sure the **Computation Method** is set to **Radiometry** and change the **Approximate number of vertices** to 20.

Fewer vertices will reduce processing time.

3. Click **Run**.

The updated scene footprints are displayed in the map. They contain only the image data, not the surrounding **NoData** pixels.

4. In the **Catalog** pane, right-click the mosaic dataset, point to **Optimize**, and click **Build Overviews**.

5. Accept all the default settings in the **Build Overviews** tool and click **Run**.

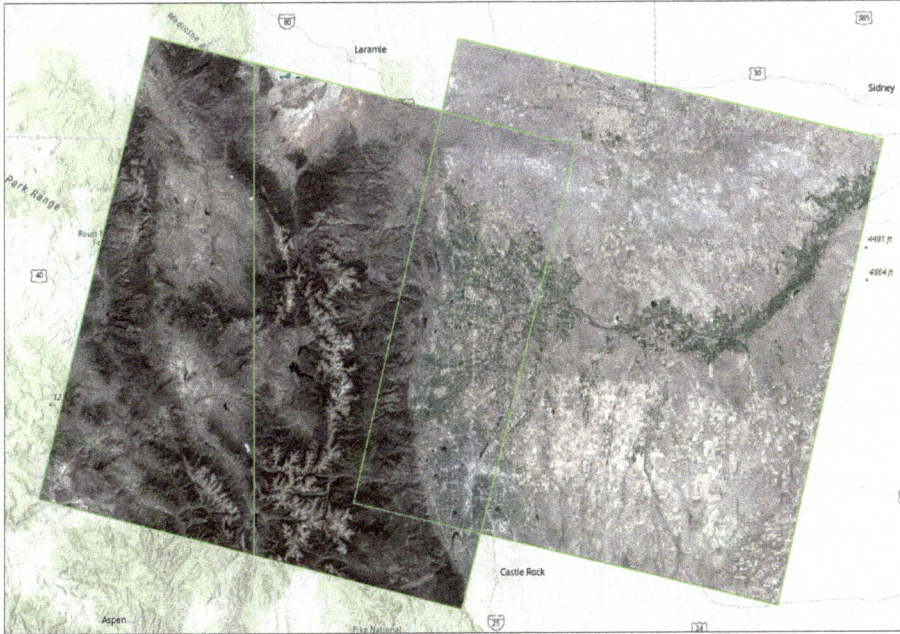

Now when you zoom far out to a small scale, the mosaicked imagery is displayed.

Tutorial 4-2: Work with the mosaic dataset

The mosaic dataset uses meta information to tailor guided workflows and properly configures settings for raster processing tools and functions to display imagery and raster data appropriately. Metadata for the cumulative mosaic dataset, as well as for individual items in the mosaic dataset, is managed, queried, and processed to intelligently display, process, and analyze collections of imagery and raster data.

You will explore the properties and functionality of the mosaic dataset in the following sections.

View mosaic dataset properties

1. In the **Catalog** pane, right-click **COfrontrangeMosaic** and click **Properties**.

    The **Mosaic Dataset Properties** window appears. The property groups for a mosaic dataset are **General**, **Defaults**, and **Manage**.

2. On the **General** tab, review the groups of properties, including **Data Source**, **Raster Information**, **Band Metadata**, **Extent**, and **Spatial Reference**.

3. Click the **Defaults** tab where the default settings to display and process the mosaic dataset are listed.

    The properties designated with an edit button (pencil) allow you to edit the options.

4. Click **OK** to close the **Mosaic Dataset Properties** window.

Use the Mosaic Layer tab

Next, you will explore visualization and enhancement of the mosaic dataset.

1. In the **Contents** pane, click the **COfrontrangeMosaic** layer.

    When the mosaic dataset is selected, the **Mosaic Layer** and **Data** tabs are available on the ribbon.

2. On the ribbon, click the **Mosaic Layer** tab, which opens the mosaic display tools.

3. In the **Rendering** group, click the **Band Combination** list to display the renderings for Landsat 8 OLI data.

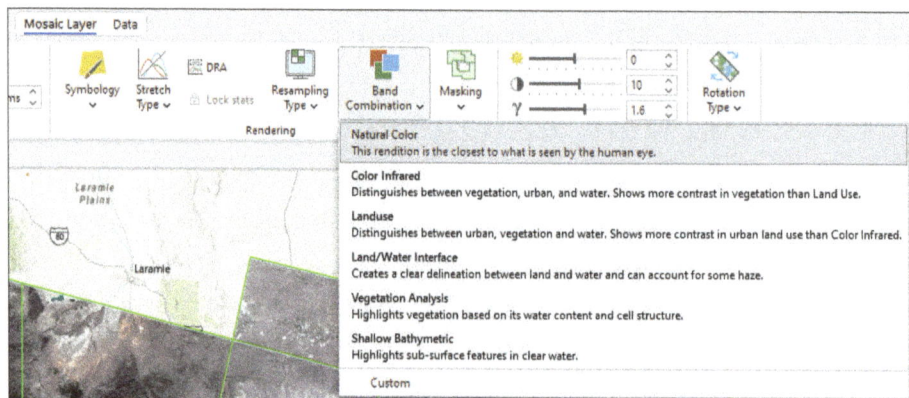

4. Click the band combinations to display information for visual analysis.

    As you change the band combination, the information associated with each visualization is updated under the image in the **Contents** pane.

5. Select the **Color Infrared** band combination.

6. In the **Rendering** group, click the **Stretch type** list, which displays the options for image histogram stretching.

7. Click the **Stretch type** options to see their effect and identify which stretch types enhance the features of interest.

8. Select the **Percent Clip** stretch type.

9. In the **Rendering** group, click **DRA**.

    Dynamic Range Adjustment (DRA) dynamically adjusts the histogram stretch depending on the content in the map display. When you zoom into a detail, the image histogram is dynamically optimized for the display extent. When you zoom out and more diverse features are displayed in the map, the histogram is readjusted to accommodate all features and conditions.

10. Pan and zoom to the mosaic layer to explore the imagery and rendering settings.

Use the Data tab

Next, you will use the **Data** tab to explore tools for managing and exploring the mosaic dataset.

1. With the **COfrontrangeMosaic** layer selected in the **Contents** pane, click the contextual ribbon **Data** tab, which opens the data management tools on the toolbar.

2. In the **Table** group, click **Attribute Table**. Use the slider bar to view all the data fields in the attribute table.

    The attribute table of the **COfrontrangeMosaic** layer appears. It contains metadata information for each individual item in the mosaic dataset. The Landsat imagery, **OBJECTID 1** and **2**, lists data for dozens of fields.

    The other items in the attribute table are mosaic dataset overviews at different resolutions.

3. Click the items in the attribute table to see them highlighted in the map. If you want to select more than one item, press **Shift** on your keyboard and click to select additional items.

4. Click the tools on the **Data** toolbar, such as **Operations**, **Sort**, and **Resolve Overlap**, and explore the options.

Use Lock Raster

Display and image processing operations can be performed on the entire mosaic dataset, a specific image contained in the mosaic dataset, or groups of images.

You will use the **Lock Raster** functionality to filter specific mosaic dataset items and process them uniquely.

1. In the attribute table, click **OBJECTID 1**.

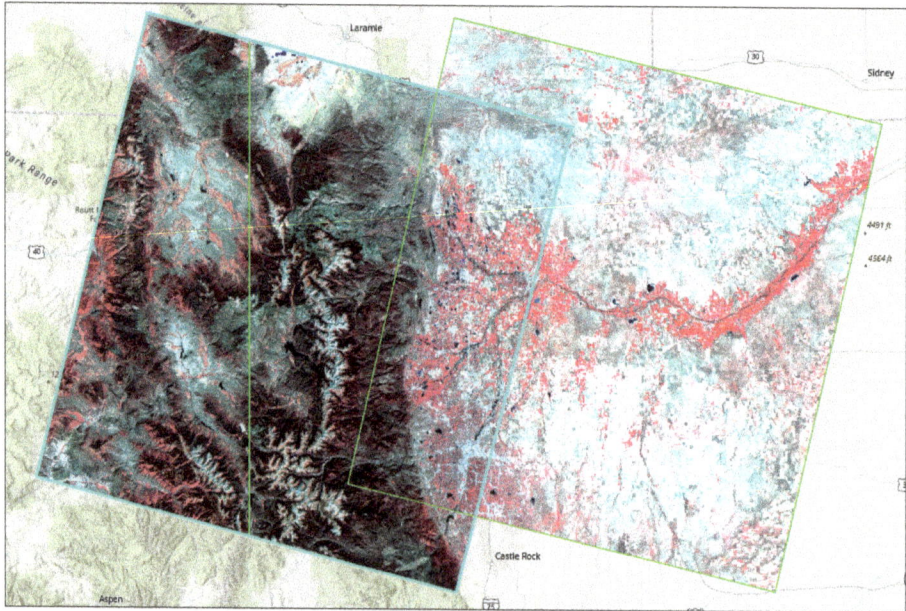

The corresponding Landsat scene is highlighted in the map.

2. On the **Data** tab, in the **Selection** group, click the **Operations** list and click **Lock To Selection**.

    Only **OBJECTID 1** (LC08_L2SP_034032_20220926_20221004_02_T1.txt) will be displayed in the map.

3. In the **Processing** group, click the **Processing Templates** list and then click **Add custom**.

4. In the **Select Raster Function Template** window, expand the project folder and select the **Raster Functions** folder. Click to select **BurnComposite.rft.xml** and click **OK**.

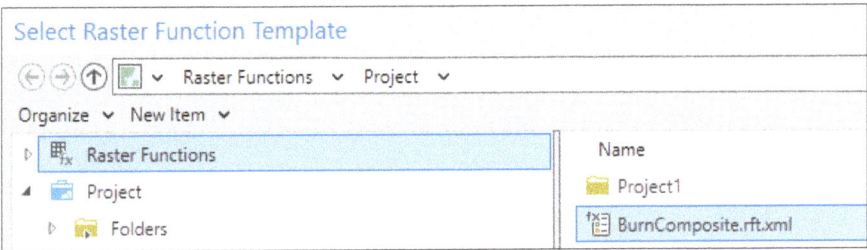

5. Open the **Processing Template** list to confirm the **BurnComposite** raster function template is applied to the **OBJECTID 1** image.

6. On the **Mosaic Layer** tab, in the **Rendering** group, click **DRA**.

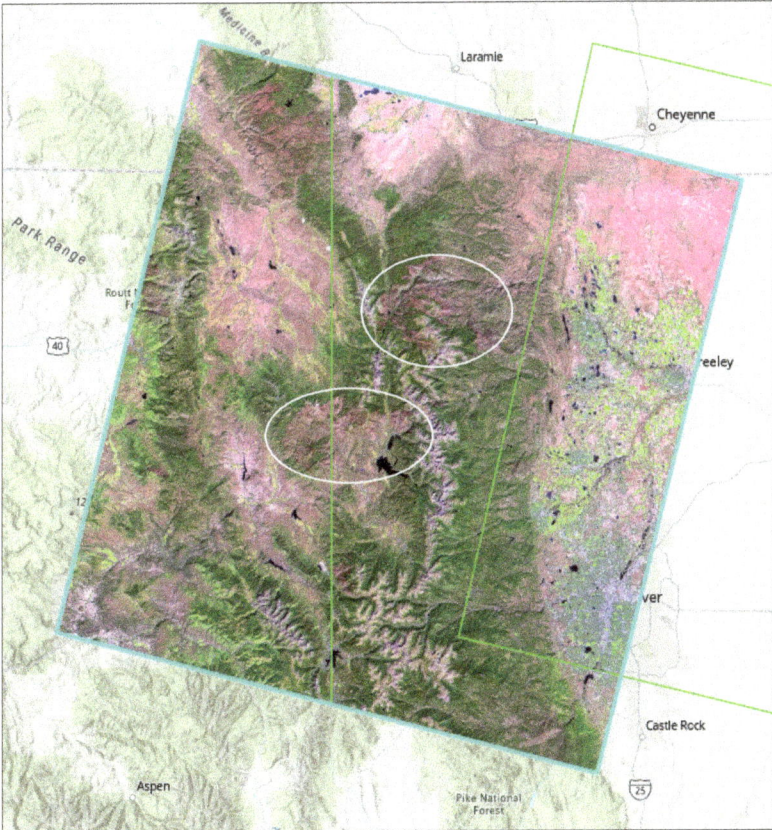

The Landsat image was rendered to enhance two burned areas indicated by a mauve color, circled in white in the figure. The band combination is indicated for the image as **ShortwaveInfrared_1** displayed as red, **NearInfrared** displayed as green, and **Green** displayed as blue.

7. On the **Data** tab, in the **Image Display Order** group, click the **Sort** list and select **North-West**.

Both Landsat images are visible on the map now.

Additional functionality

The mosaic dataset is a robust data model based on the file geodatabase structure and used to manage, catalog, visualize, and serve collections of image and raster data as a single, seamless image or surface. Functionality to facilitate these operations is briefly described in the following sections.

Raster Item Explorer pane

The Raster Item Explorer option can be used to filter items from a large collection of imagery in either a mosaic dataset or an image service and to explore the properties of individual items, add them to a map or scene, and view and edit the processing applied to an item.

In the Raster Item Explorer pane, click the Select tab, filter the list of items listed in the pane, and select one or more mosaic dataset items. The selected items can then be visualized and processed as a group. Click the Inspect tab to display a preview of the item and review details about the image item.

Seamlines

You can use seamlines to define the edges along which the images in the mosaic dataset are mosaicked together. Using seamlines removes overlap between images in the mosaic dataset and provides flexibility to adjust the cut lines between images to aid in creating a "seamless" mosaic. For example, you can define a seamline along a street or other human-made or natural features. Specify whether you want the boundary between images to be joined end to end (no overlap) or define a blend width value and type that will occur between pixels in overlapping images.

To create seamlines for a mosaic dataset, use the Build Seamlines tool. Seamlines are similar to footprints, in which one polygon represents each image. The shape of the polygon represents the part of the image that is used to generate the mosaicked image when viewing the mosaic dataset. Once the seamlines are built, a Seamline layer is present in the Contents pane each time you add the mosaic dataset to the map.

Sort using mosaic methods

You can specify how you want the imagery in your mosaic dataset to be prioritized and ordered for viewing. The By Attribute sort method offers flexibility to sort by specific attributes, including by time and date.

Resolve overlapping pixels

After you've determined the method for ordering, you can fine-tune the results. Use the Resolve Overlap button to access a set of mosaic operators.

Color balancing

Use the Color Balance Mosaic Dataset tool to adjust the appearance of images in the mosaic dataset based on the statistics and histogram of each image. Define a target image that all the other images will be matched to, specify a color surface used for the adjustment, and choose a color balance method, such as Dodging, Global Fit, Histogram, and Standard deviation.

Export Raster

You can download the entire mosaic dataset as a single image file, or you can specify individual images or group of images from the attribute table to be downloaded as image files stored on disk. In the Tools group, click the Export Raster button to open the Export Raster window. Fill out the dialog box according to your requirements and click Export to save the mosaic image layers to disk.

Summary

In this chapter, you learned how to create a mosaic dataset using Landsat 8 images. You then learned how to display and enhance the mosaic dataset, including how to select specific mosaic dataset items for processing and display. The skills and techniques included in this tutorial will be useful whenever you need to display, process, and manage collections of imagery for a wide variety of applications, including multidimensional data analysis, photogrammetric processing, multispectral and hyperspectral image analysis, and more.

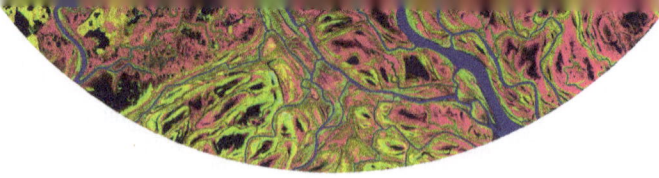

CHAPTER 5
Performing radiometric calibration

Objectives

- Use the Apparent Reflectance function to perform atmospheric calibration.
- Use a formula to convert the radiometric values of an image.

Introduction

Satellite-based sensors collect all reflected energy through the atmosphere, which is in a state of constant flux. The atmosphere scatters and absorbs all this solar radiation as well as emitted energy from the surface and therefore affects the energy that reaches the sensor and is recorded. Ideally, researchers want to correct for these atmospheric conditions. However, to perform atmospheric correction, you would need complex atmospheric models or information about atmospheric conditions at the time of collection. ArcGIS Pro provides a way to account for some of these conditions and allows you to radiometrically calibrate imagery for common, well-known atmospheric effects. Radiometric calibration—in the form of atmospheric calibration—rescales your imagery and provides scientists and analysts with a more accurate measurement of surface properties. Additionally, atmospheric calibration ensures a consistent measurement scale that can be understood and applied to different images collected at different times and through varying atmospheric conditions.

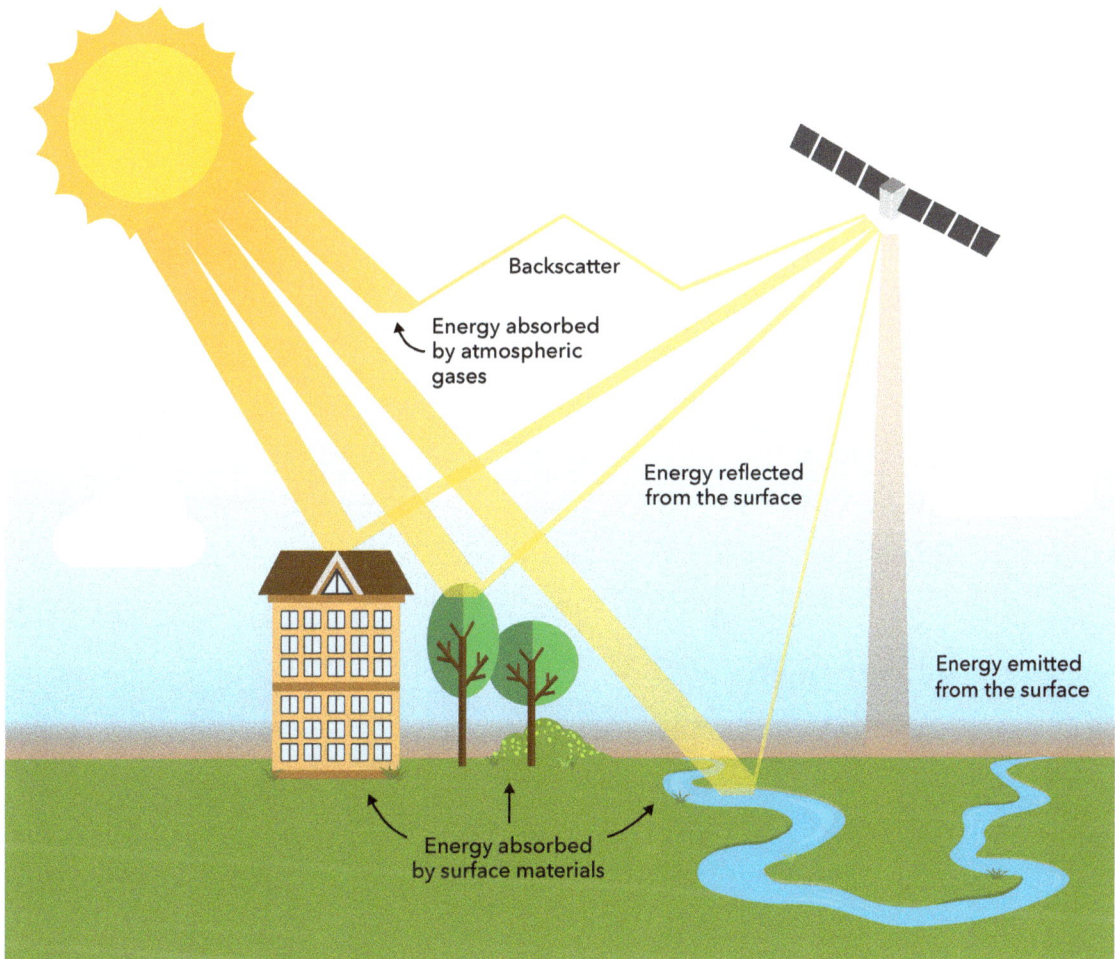

Figure 5-1. Satellite sensors collect reflected energy through a constantly changing atmosphere. Raster functions can be used to radiometrically calibrate images for use in different types of analysis, including change detection or image classification.

In this chapter, you'll learn how to perform this type of radiometric calibration of imagery. You'll use two raster functions to calibrate Landsat 8 imagery to surface reflectance values for use in further analysis. As you learned previously, imagery is often delivered as uncalibrated digital number (DN) values. Although imagery can be used and analyzed using this baseline radiometric resolution, for many scientific applications it is best to calibrate the image to a standardized radiometric resolution.

Tutorial 5-1: Perform atmospheric calibration on an image

In this tutorial, you'll use the **Apparent Reflectance** function to calibrate a Landsat 8 image to top of atmosphere reflectance values. ArcGIS Pro reads key metadata from sensor information when loading a raster product, making it easy to use prebuilt raster functions. The **Apparent Reflectance** function uses published formulas to convert DN values to radiance values and then to reflectance, or albedo, values. To use this function, your image requires specific metadata:

- Acquisition date and sun elevation for the dataset
- Radiance gain, radiance bias, and sun irradiance for each band
- Reflectance gain and reflectance bias (for Landsat 8)

The **Apparent Reflectance** function can be used only with specific imagery (See "Information at Your Fingertips" at the end of this chapter).

Download the tutorial data and set up the project

1. Go to links.esri.com/Imagery20Data and download the data for chapter 5.

2. Unzip the folder to **C:\Top20Imagery**.

    > **Note:** In the second chapter, you created a folder named **Top20Imagery** on your C: drive. If you haven't done that, create that folder now. Now and in subsequent chapters, you will download and unzip the data for each chapter to this folder.

3. Inside the **Top20Imagery_05** folder, double-click **Top20Imagery_05.aprx** to open the ArcGIS Pro project for this chapter.

4. In the **Catalog** pane, expand **Folders** > **Top20Imagery_05** and then expand the **2025_0328-233_078_L1TP (Chile)** folder.

    > *Important: Make sure you select the Level-1 Precision and Terrain Correction (L1TP) folder.*

This folder contains a Landsat 8 Operational Land Imager/Thermal Infrared Sensor Level-1, Collection 2 Precision and Terrain Correction Product (L1TP)[1] of the northern border between Chile and Argentina. This type of product has calibrated and scaled DN values. In chapter 2, you learned how the DN values relate to collected and measured intensity values from features on the ground.

In ArcGIS Pro, these values are represented as 16-bit unsigned data. This is the radiometric resolution of the image, the recorded and represented brightness of the image pixel values.

5. In the **Catalog** pane, right-click the **LC08_L1TP_233078_20250328_20250401_02_T1_MTL.txt** raster product and click **Add To Current Map**.

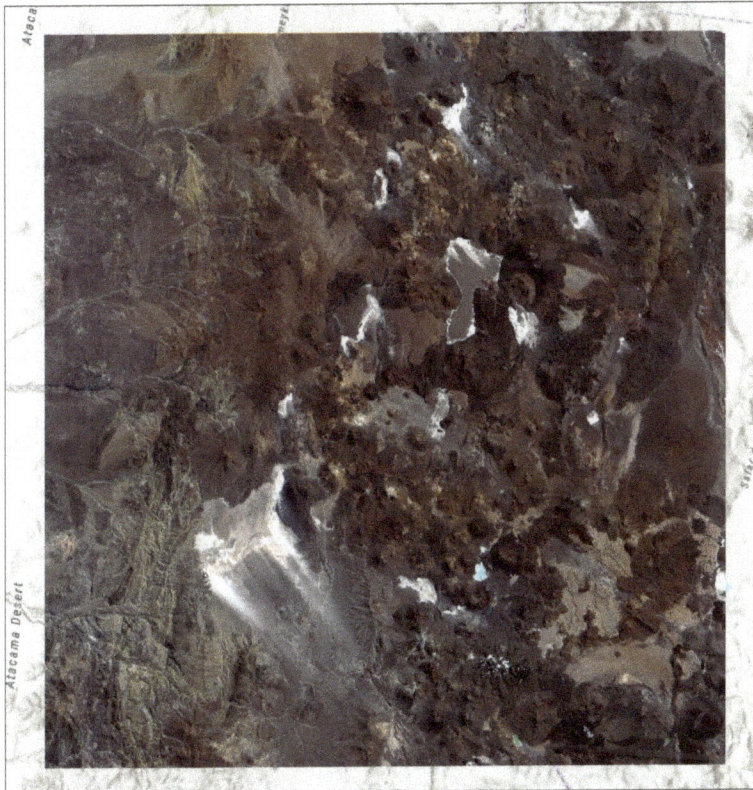

The new image appears in the **Contents** pane and on the map.

6. Using skills you've already learned, explore the image.

7. In the **Contents** pane, right-click the **Multispectral_LC08_L1TP_233078_20250328_20250401_02_T1_MTL** layer and click **Zoom To Layer**.

Several collection tiers are available from USGS. This image is a Tier 1 product. Tier 1 images have been corrected by USGS and contain well-characterized radiometry—that is, data values across the various multispectral bands. One advantage to this processing level is that you can use formulas and internal metadata related to the sensor settings to calibrate the image for some atmospheric effects.

Calibrate an image to apparent reflectance

You can radiometrically calibrate this image to reflectance values by using a raster function.

1. On the **Imagery** tab, in the **Analysis** group, click **Raster Functions** to open the **Raster Functions** pane.

2. At the top of the **Raster Functions** pane, in the search bar, type Reflectance.

3. Under **Correction**, click the **Apparent Reflectance** function.

    The **Apparent Reflectance** function is used to adjust the brightness values of some satellite imagery based on the scene illumination and sensor-gain settings. The calibration function uses sun elevation, acquisition date, and sensor gain and bias for each band to derive top of atmosphere reflectance from the DN values of imagery product types. The images are adjusted to a theoretically common illumination condition. One result of this adjustment is that there should be less variation between scenes from different dates and different sensors. This type of calibration is especially useful for analysis, such as image classification, indexes and ratios, and change detection.

    All the information required for the correction is extracted from the key metadata properties for each image when the function is initialized. If these metadata properties are present, selecting the image layer populates these values automatically.

4. In the **Raster Functions** pane, for **Raster**, select the **Multispectral_LC08_L1TP_233078_20250328_20250401_02_T1_MTL** layer.

    All the relevant information for **Reflectance Gain** and **Bias** values is added to the table for each band. In addition, the **Sun Elevation** in degrees is automatically added. The appropriate **Scale Factor** and **Offset** are similarly populated from the metadata.

5. Under **Sun Elevation (degrees)**, check the box for **Albedo**.

    You need to check the **Albedo** box to ensure that your calibration returns reflectance values. These values represent a brightness layer based on the proportion of radiation that is reflected by the surface. Albedo is expressed as a dimensionless 32-bit floating point number between 0 and 1. Zero corresponds to a black body that absorbs all incident radiation, such as wet coal, and 1 corresponds to a body that reflects all incident radiation, similar to fresh snow. These values can be thought of as a percentage of reflectance where zero is total absorption and 1 is 100 percent reflectance.

> **Note:** Once calibrated, there are two ways to express radiometric signal return for imaging sensors: radiance values or reflectance values. For most remote sensing analysis, reflectance values are preferred and are used by scientific users for complex modeling and technical remote sensing applications.

6. At the bottom of the pane, click **Create new layer**.

The new result layer is visible in the **Contents** pane and added to the map. This function modifies the image values—the radiometric resolution of the image—so previous statistics and histograms are no longer valid. As a result, a new symbology rendering is applied so the image may appear slightly darker.

Review the corrected image

1. On the **Map** tab, in the **Navigate** group, click **Explore** and select **Visible Layers**.

2. Click in the map to select a pixel.

Note: Your pixel values will differ from those shown, based on where you clicked in the image.

The new image is now radiometrically calibrated to a known standard: reflectance values.

3. In the **Contents** pane, uncheck the boxes for the **Apparent Reflectance_** and **Multispectral_ LC08_L1TP_233078_20250328_20250401_02_T1_MTL** layers to turn off their visibility.

Tutorial 5-2: Radiometrically calibrate an image using the Calculator function

In this tutorial, you'll use a different raster function to radiometrically calibrate an image already corrected to surface reflectance by USGS but not provided with surface reflectance values—that is, values between 0 and 1. First, you'll add a Landsat 8 Level-2 Science Product (L2SP)[2] of the same area in northern Chile.

Add a new image to your map

You'll now add a new folder containing a Landsat 8 Level-2, Collection 2 Science Product (L2SP) of the same area.

1. In the **Catalog** pane, collapse the **2025_0328-233_078_L1TP (Chile)** folder and expand the **2025_0328-233_078_L2SP (Chile)** folder.

   > *Important:* *Make sure you select the Level-2 Science Products (L2SP) folder.*

   Like the previous folder, this folder contains a Landsat 8 image of the same geographic area you saw in the previous tutorial. However, this image has been processed as a Landsat 8 Operational Land Imager/Thermal Infrared Sensor Level-2, Collection 2 Science Product (L2SP) of surface reflectance. This type of image is, as the name suggests, a science product, and as such, you can use it for several scientific applications needing surface reflectance or surface temperature values. The data is processed and saved by USGS as scaled integers (16-bit un-signed data) from the original floating point (32-bit) data. By converting to 16-bit data, USGS saves disk space and allows for faster download times from its data delivery sites. To perform this conversion, each pixel has an offset value applied and then is multiplied by a gain value to set them into the 16-bit integer data range (0–65,535).

2. In the **Catalog** pane, right-click the **LC08_L2SP_233078_20250328_20250401_02_T1_MTL.txt** raster product and click **Add To Current Map**.

You'll notice that the image appears discolored. This is because the surface reflectance bands need to be adjusted in the **Contents** pane to replicate a natural color appearance.

3.　In the **Contents** pane, under the **Surface Reflectance_LC08_L2SP_233078_ 20250328_20250401_02_T1_MTL** layer, right-click the **Red** channel and click **sr_band4**.

4.　Repeat this process so that the rendered band combination is as follows:
- **Red**: sr_band4
- **Green**: sr_band3
- **Blue**: sr_band2

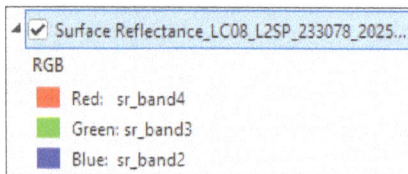

Now that you've loaded the new image and set the rendering to natural color, you can rescale the image using a set formula to radiometrically calibrate the image, meaning that you can convert it back to its original reflectance values as computed by USGS.

Use a raster function to calculate a new radiometric range

You can use the **Calculator** raster function to convert the 16-bit unsigned image into the calculated 32-bit floating point values derived by NASA and USGS.

1.　Open the **Raster Functions** pane. In the search field, type Calculator.

2.　Under **Math**, click the **Calculator** function.

USGS publishes the values and formula necessary to rescale the delivered 16-bit imagery to the actual surface reflectance values. To convert the 16-bit imagery, you must apply a scale factor (or multiple) to each pixel in each band. This scale factor value is 0.0000275. There is an additional additive offset that also must be used. This value is −0.2. The formula you must apply to all pixels is the following:

$$\text{Surface Reflectance} = DN \times 0.0000275 + -0.2$$

You can use the calculator to build this formula and apply it to the image. To make the calculator easier to use, you'll first set the surface reflectance layer as a raster variable for the input DN values.

3. In the **Raster Functions** pane, for **Raster Variables**, type DN in the first field. In the second field, select the **Surface Reflectance_LC08_L2SP_233078_20250328_20250401_02_T1_MTL** layer.

4. Click inside the **Expression** field to start adding your formula. Type (DN*0.0000275) + -0.2.

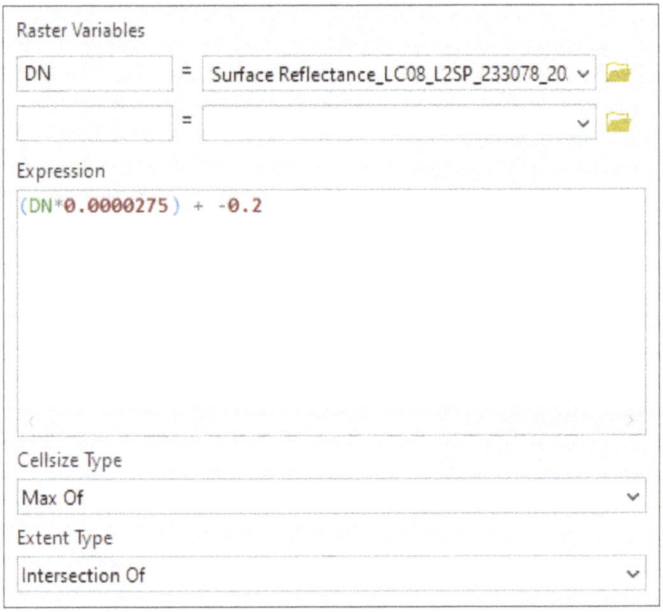

Raster Variables

| DN | = | Surface Reflectance_LC08_L2SP_233078_20 ⌄ | 📁 |
| | = | ⌄ | 📁 |

Expression

(DN*0.0000275) + -0.2

Cellsize Type

Max Of ⌄

Extent Type

Intersection Of ⌄

5. Click **Create new layer**.

6. In the **Contents** pane, change the band combination to the natural color rendering.

> **Remind me how:** Right-click the appropriate color channel and then click the desired band.
> - Red: sr_band4
> - Green: sr_band3
> - Blue: sr_band2

7. On the **Map** tab, in the **Navigate** group, click **Explore**.

8. Click in the map to select a pixel.

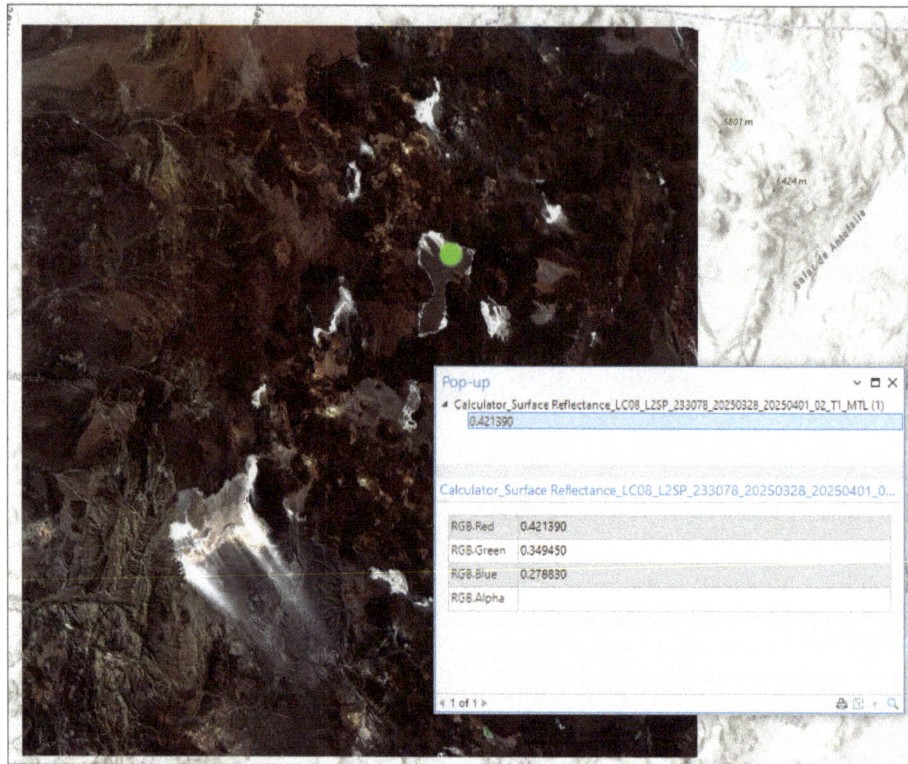

The new raster function result now represents the correct surface reflectance values calculated by USGS for use in scientific processing. You can also compare the two results.

9. In the **Contents** pane, turn off the **Surface Reflectance_LC08_L2SP_233078_20250328_20250401_02_T1_MTL** layer and turn on the **Apparent Reflectance_Multispectral_LC08_L1TP_233078_20250328_20250401_02_T1_MTL** layer result.

10. Click in the map to select a pixel.

The pop-up shows results from both layers that have their visibility turned on. There will be an expected variance between the two results. However, either result will be valid for processing. Remember that one image, the **Apparent Reflectance** result, is a top of atmosphere calibration of radiometric values whereas the **Calculator** result is the surface reflectance as derived and calculated by USGS. If your imagery provider or sensor publishes rescale values and offsets, you can use the same **Calculator** function to adjust and modify your image's radiometric values.

Summary

In this chapter, you used two raster functions to radiometrically calibrate two types of Landsat imagery. You used the Apparent Reflectance function to calibrate an image to top of atmosphere reflectance values. You also used the Calculator raster function to rescale and radiometrically calibrate a 16-bit unsigned image to 32-bit floating point values representing surface reflectance values as computed by the USGS Landsat program.

Information at your fingertips

Additional information about the Apparent Reflectance function

The function performs two corrections. The first is based on the gain settings. The original brightness values are re-created from the image values by reversing the gain equations. This converts the image to radiance values.

The second correction is for differences in sun angle and brightness. This correction converts the radiance values to top of atmosphere reflectance values.

The original brightness values are adjusted to a common lighting condition by normalizing scenes captured under variable illumination conditions. In general, although the output image data type is the same as the input image data type, the output values are lower than the input values and are clipped to the valid data range.

This function can be used only with specific imagery. The applicable sensors are the following:

- Landsat MSS
- Landsat TM
- Landsat ETM+
- Landsat 8
- IKONOS
- QuickBird
- GeoEye-1
- RapidEye
- DMCii
- WorldView-1

- WorldView-2
- SPOT 6
- Pleiades

Note: Detailed formulas for these two calibrations can be found in the *Landsat 8 Data Users Handbook*, November 27, 2019 (Landsat Missions/Manuals, US Geological Survey). This resource also contains useful information on the Landsat program, radiometry, instrumentation, engineering, and other reference material.

Resources about radiometric calibration

Publicly available resources on the Landsat program provide a wealth of information to help scientists and researchers. Visit links.esri.com/ImageryResources to learn more.

Notes

1. Earth Resources Observation and Science (EROS) Center. 2020. Landsat 8–9 Operational Land Imager/Thermal Infrared Sensor Level-1, Collection 2 (dataset), US Geological Survey. https://doi.org/10.5066/P975CC9B.
2. Earth Resources Observation and Science Center. 2020. Landsat 8–9 Operational Land Imager/Thermal Infrared Sensor Level-2, Collection 2 (dataset), US Geological Survey. https://doi.org/10.5066/P9OGBGM6.

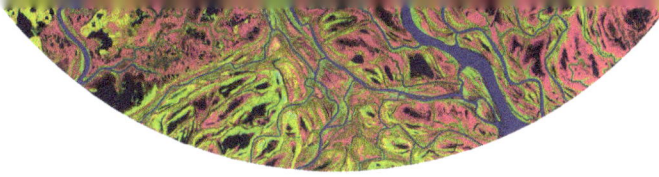

CHAPTER 6
Georeferencing imagery

Simon Woo

Objectives

- Align imagery to your GIS data or basemap.
- Review ground control point pairs.
- Save georeferenced imagery data.

Introduction

Modern satellite and aerial imagery tend to have relatively accurate location information but may need adjustments to line up with your GIS data. Georeferencing can be used to make these adjustments to align your imagery. ArcGIS Pro can perform both tasks so that your imagery aligns with your other data, and then it can be used for more accurate analysis and visualization.

Tutorial 6-1: Explore and georeference your imagery

Briefly explore your imagery data. Here you will see that your satellite imagery is already in the correct general position. Upon further inspection, you will notice an offset between your imagery and the other data, including the basemap.

Download the tutorial data and set up the project

1. Go to links.esri.com/Imagery20Data and download the data for chapter 6.

2. Unzip the folder to **C:\Top20Imagery**.

   > **Note:** In the second chapter, you created a folder named **Top20Imagery** on your C: drive. If you haven't done that, create that folder now. Now and in subsequent chapters, you will download and unzip the data for each chapter to this folder.

3. Inside the **Top20Imagery_06** folder, double-click **Top20Imagery_06.aprx** to open the ArcGIS Pro project for this chapter.

Explore the map

1. Pan and zoom around the map.

2. In the **Contents** pane, select the **Gentilly** raster layer.

3. On the ribbon, click the **Raster Layer** tab. In the **Compare** group, use the **Swipe** tool to reveal what is under the raster layer and compare it with the basemap.

4. On the **Map** tab, in the **Navigate** group, click **Bookmarks** and double-click the bookmark named **Check Offset**.

   The bookmark shows an area in the northwest portion of the image.

5. Turn the **Gentilly** layer on and off and notice that the **edge of pavement** layer is not aligned with the **Gentilly** layer but is fairly aligned with the basemap.

Satellite imagery ©2018 Vantor.

A basemap is used only as a background image and cannot be analyzed or altered, so you need to make sure the **Gentilly** layer is aligned with your data. Next, you will rectify this image so that it aligns better with your other GIS data. This is accomplished using tools available on the **Georeference** tab.

Create your first ground control point

Tools on the **Georeferencing** tab are used to align your imagery source layer to the basemap. This ensures that all your data is aligned before you can properly analyze or visualize your spatial data. To align the data, you will need to create ground control points (GCPs). The points have a "from" and "to" location—the "from" point is the location on the image you need to georeference, and the "to" point is the correct position on the target reference layers. In this case, the **Gentilly** layer is your source layer that you want to georeference. The target layers will be the imagery basemap and other GIS layers that may be available.

> **Note:** When georeferencing your imagery, you must choose features on the ground (not elevated features) that will not move over time. Human-made features, such as the base of utility poles, concrete intersections, and building footprints, are good choices.

1. In the **Contents** pane, turn on visibility for the **Gentilly** layer.

2. On the **Imagery** tab, in the **Alignment** group, click the **Georeference** button.

   The **Georeference** tab appears.

   When you create control points, you should create them in each corner of the image so that all four corners of the image are accounted for. Sometimes this is not possible because of obstacles in the imagery, such as water, clouds, or other areas where you are not able to find control point pairs. Once you have covered the four corners as well as possible, you will add more points, spread out throughout the image.

3. On the **Map** tab, click **Bookmarks** and select the **NE** bookmark.

Satellite imagery ©2018 Vantor.

The map zooms in on a tennis court located in the northeast part of the image.

4.   On the **Georeference** tab, click **Add Control Points**.

5.   Make sure the **Gentilly** layer is selected. On the map, click the bottom left corner of the right-side tennis court.

This will be the source point.

Satellite imagery ©2018 Vantor.

6. Turn off visibility for the **Gentilly** layer.

>   **Tip:** While georeferencing, press the *L* shortcut key to toggle the visibility of the layer you are georeferencing.

7. Click on the same location shown in the basemap.

    This will be the target point.

Satellite imagery ©2018 Vantor.

8. Turn the visibility of the **Gentilly** layer back on.

    As you create each GCP, the image automatically shifts so that you can see the updates.

Create more GCPs

You will create more GCPs using the same pattern—placing a source point on the **Gentilly** layer and pointing it to a target point on the basemap layer.

1. In the list of **Bookmarks**, select **NW**. Use the top corner of the pool to create a second control point.

2. Select the **SW** bookmark. Use a corner of the pool to create a third control point.

3. Select the **SE** bookmark. Use the inside corner of the sidewalk to create a fourth control point.

4. Select the **Central** bookmark. Use a corner of the court for a fifth control point.

5. Select the **GCP 6** bookmark. Zoom to one of the poles and use the bottom of the pole to create a sixth control point.

6. Select the **GCP 7** bookmark. Use one of the corners of the sidewalk intersection to create a seventh control point.

7. On the **Georeference** tab, click **Transformation** and then click **Second Order Polynomial**.

A second-order polynomial transformation requires a minimum of six GCPs. It is usually used when there is distortion in the image. Because our data has a lot of obliqueness to it, we will use the second-order polynomial transformation. The total root mean square (RMS) error for a second-order polynomial is calculated only after seven GCPs have been created.

You have now created a GCP in each of the four corners, plus a few additional points. You should always make sure that there are GCPs covering the image evenly. If there are any areas without GCPs nearby, use the steps here to create a new GCP in that area. Once you have decent coverage of GCPs throughout the image, you are ready to review your points and your accuracy.

On your own

Continue to create CGPs so that there is even coverage throughout the image. You can use the bookmarks, and the notes within each bookmark, to help you create more points.

Tutorial 6-2: Review the control points and create a georeferenced image

Once you have collected your GCPs, you can now review your points. There's no rule on how many GCPs should be created for your georeferencing task, as long as you have the minimum number of points needed for the transformation type.

Note: The transformed image is displayed in the map as a dynamic image layer, where the image transformation is computed on the fly and displayed as you pan and zoom the image layer.

To review each GCP, you can use the **Control Point Table** and the map display to help you evaluate each point.

Verify the accuracy of your GCPs

1. On the **Georeference** tab, in the **Review** group, click **Control Point Table**.

    The **Control Point Table** appears, docked below the map display.

	Link	Source X	Source Y
☑	1	3,688,314.900837	556,944.360922
☑	2	3,677,954.706638	556,971.268856
☑	3	3,678,078.837961	549,318.408029

New Orleans: Gentilly ✕

2nd Order Polynomial ⌄

2. In the **Control Point Table**, double-click the gray area on the left of the first check box.

    This zooms to the first control point in the table.

    > **Tip:** If you are not zoomed in close enough to examine the accuracy of the point, you can click the **Zoom to Selected** button in the table until you are at a proper magnification.

3. Examine the GCP to make sure that the **From** and **To** points match the same location accurately.

4. Repeat this process to check the accuracy of each point. If a control point looks inaccurate, in the table, click the **Delete Selected** button to delete any inaccurate GCPs. If you delete any inaccurate GCPs, you will need to create another GCP in its place to make sure you still have enough GCPs.

    > **Tip:** You can also readjust a point using the **Select** tool in the **Review** group of the **Georeference** tab.

Once you have reviewed each of the control points, you need to make sure that the RMS error is acceptable. An RMS error should be less than the size of a pixel. In this case, the data has a pixel size of 1 (foot); therefore, you want an RMS error less than 1.

Note: Although the RMS error is a good assessment of the transformation's accuracy, don't confuse a low RMS error with an accurate registration. This is why you need to review each point within the table.

5. On the map, look at the top right corner and locate the **Georeferencing** heads-up display (HUD). Make sure that the RMS error for **Forward** and **Reverse** are both less than 1.

Georeferencing: Gentilly

Transformation: 2nd Order Polynomial
Controls Points: 7 / 7
Total RMS Errors
 Forward: 0.441854
 Inverse: 0.441830
 Forward-Inverse: 0.000443

Note: Your numbers may differ. If the total RMS error is under 1, you can continue.

6. Compare your **Gentilly** layer against the **edge of pavement** layer. It now aligns much better than it did previously.

Satellite imagery ©2018 Vantor.

Because you now have accurate GCPs that cover your image, you are ready to save the results.

Save your results

You can save the georeferencing information back to the current file, or you can create a new raster dataset. In this case, you will save the georeferencing information to the current TIFF file.

1. On the **Georeference** tab, in the **Save** group, click **Export Control Points**.

2. Save the resulting file to **Project > Folders > Top20Imagery_06 > GCP** and name it ControlPoints.

   This saves the control points you created to a text file. It is a good idea to save these control points, especially since you will be creating a new raster dataset.

3. In the **Save** group, click **Save**.

   Once the raster dataset has been saved, you can close the georeferencing session.

4. In the **Close** group, click **Close Georeference**.

Summary

In this chapter, you viewed a project with a raster dataset. At first, the raster looked as if it was in the correct geographic location. However, upon closer inspection, you noticed that it needed a slight adjustment. You aligned the raster using the Georeferencing tab, so that your image was better aligned with the imagery basemap and other GIS data. You learned how to review the GCPs, both the point location and the total RMS error. You saved your GCPs to a text file and updated your georeferenced results. Your imagery is now prepared for use in GIS tasks.

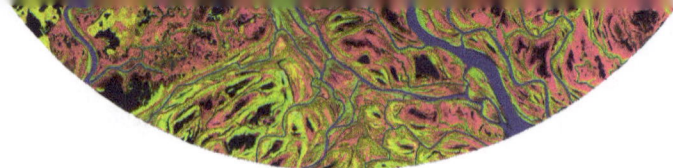

CHAPTER 7
Creating photogrammetric products

Christopher Patterson and Jeff Liedtke

Objectives

- Create an ArcGIS Reality mapping workspace to ingest and manage a collection of aerial images.
- Set up Reality mapping project settings.
- Perform a block adjustment.
- Review tie point residuals and accuracy reporting in the logs file.
- Review the Adjustment Report.
- Generate a DSM, True Ortho, and DSM mesh.

Important: To complete this chapter, you must have the ArcGIS Reality extension for ArcGIS Pro installed. For more information, visit links.esri.com/reality_setup. To quickly see if you have the Reality extension for ArcGIS Pro installed, on the main ArcGIS Pro ribbon, click the Imagery tab. On the left, you should see a Reality Mapping group.

If Reality mapping is not installed, contact your system administrator to obtain the installers and license keys.

In tutorial 7-3, the product generation may take a few hours, depending on the configuration and resources of your computer.

In ArcGIS Pro, you can correct digital aerial imagery collected by a professional mapping camera using photogrammetry to remove geometric distortions caused by the sensor and correct for terrain displacement. After correcting these effects, you can generate Reality mapping products.

In this chapter, you will set up a Reality mapping workspace to manage an aerial imagery collection, perform a block adjustment, and review the results. Then you'll generate digital surface model (DSM), true ortho, and 2D DSM mesh products.

Tutorial 7-1: Create a Reality mapping workspace

A Reality mapping workspace is an ArcGIS Pro subproject that is dedicated to Reality mapping workflows. It is a container in an ArcGIS Pro project folder that stores the resources and derived files that belong to a single image collection.

Download the tutorial data and set up the project

1. Go to links.esri.com/Imagery20Data and download the data for chapter 7.

2. Unzip the folder to **C:\Top20Imagery**.

> **Note:** In the second chapter, you created a folder named **Top20Imagery** on your C: drive. If you haven't done that, create that folder now. Now and in subsequent chapters, you will download and unzip the data for each chapter to this folder.

3. Inside the **Top20Imagery_07** folder, double-click **Top20Imagery_07.aprx** to open the ArcGIS Pro project for this chapter.

   A collection of digital aerial images is provided for this tutorial along with frame and camera tables.

Create the workspace

To create a Reality mapping workspace, complete the following steps.

1. On the ribbon, click the **Imagery** tab. In the **Reality Mapping** group, click the **New Workspace** arrow and then click **New Workspace**.

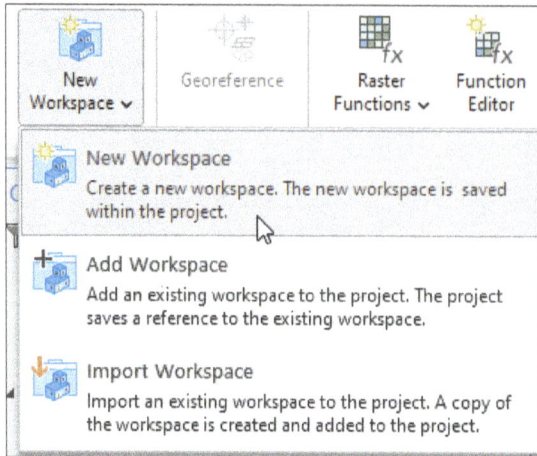

The **New Reality Mapping Workspace** pane appears.

2. In the **Workspace Configuration** step, confirm or apply the following settings:
   - **Name**: RealityMappingAerial_2D
   - **Workspace Type**: Reality Mapping
   - **Description**: The essential skills for reality mapping.
   - **Sensor Data Type**: Aerial - Digital
   - **Scenario Type**: Nadir
   - **Basemap**: Topographic
   - **Parallel Processing Factor**: 80%

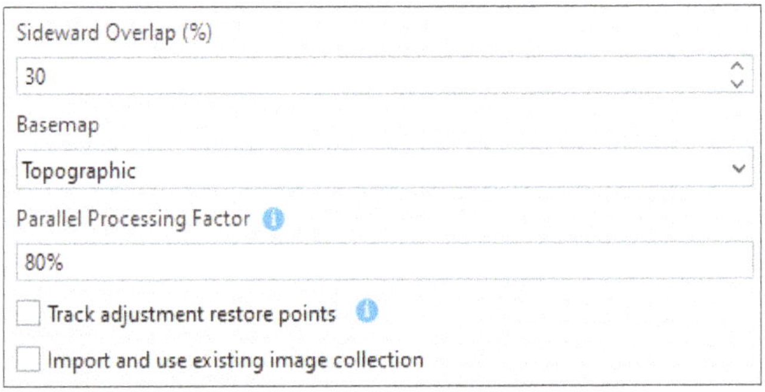

Sideward Overlap (%)

30

Basemap

Topographic

Parallel Processing Factor ⓘ

80%

☐ Track adjustment restore points ⓘ

☐ Import and use existing image collection

3. Accept all the other default values and click **Next**.

4. In the **Image Collection** step, for **Exterior Orientation File/Esri Frames Table**, click the **Browse** button.

5. Browse to **C:\Top20Imagery\Top20Imagery_07** and click the **Nadir_FramesCam.csv** frames table file. Click **OK**.

   This table, which contains the frames (individual images) and cameras information, specifies settings for computing both the interior and exterior orientation for the camera and the imagery.

6. For **Cameras**, click the **Import** (down arrow) button. Browse to the project's folder and click **Nadir_FramesCam.csv**. Click **OK**.

7. For **Workspace Spatial Reference**, ensure that the spatial reference listed is **NAD_1983_2011_StatePlane_California_V_FIPS_0405 / VCS: NAVD_1988**.

Sensor Type

Generic Frame Camera ⌄

Add

⌄ Source Data 1

Exterior Orientation File / Esri Frames Table

C:\Top20Imagery\Top20Imagery_07\Nadir_FramesCam.csv 📁

Frame Spatial Reference

NAD_1983_2011_StatePlane_California_V_FIPS_0405 / VCS: NAVD_1988 🌐

Cameras + ↓ →

✓ 0

Workspace Spatial Reference ⓘ

NAD_1983_2011_StatePlane_California_V_FIPS_0405 / VCS: NAVD_1988 🌐

Coordinate System Transformations

✏

☐ Images in this collection are adjusted

8. Accept the other default values and click **Next**.

9. In the **Data Loader Options** step, for **DEM**, click the **Browse** button.

10. Browse to **C:\Top20Imagery\Top20Imagery_07\DEM**, click **DEM_USGS_1m.tif**, and click **OK**.

Elevation Source ⓘ

Average Elevation from DEM ⌄

DEM

C:\Top20Imagery\Top20Imagery_07\DEM\DEM_USGS_1m.tif 📁

Processing Template ⓘ

Default ⌄

> Advanced Options

11. Accept the other default values and click **Finish**.

Once you create the workspace, the images and image footprints appear. The **Reality Mapping** category has also been added to the **Contents** pane. The source imagery data and derived Reality mapping products are referenced. The **Logs** are displayed below the map.

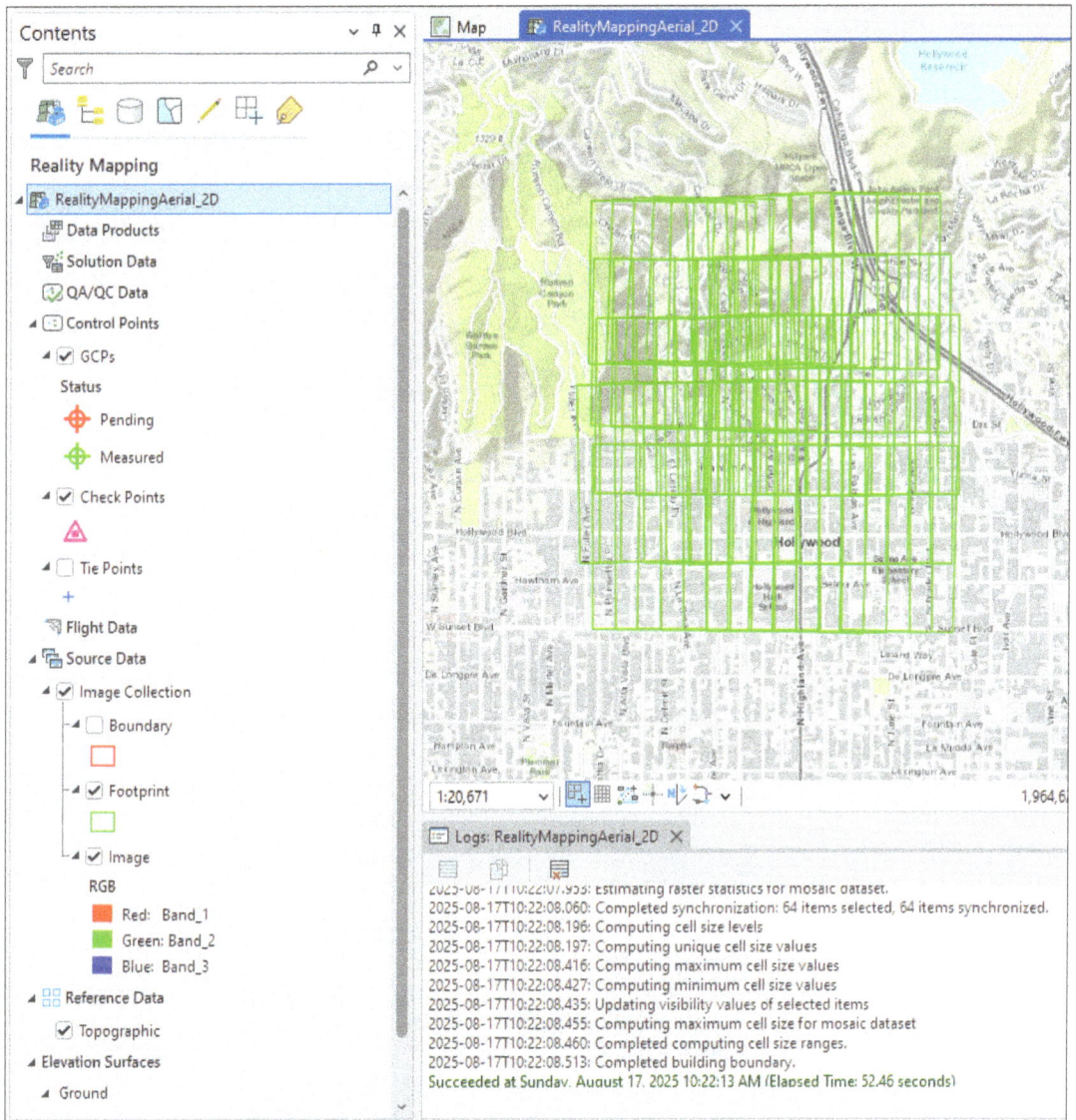

The initial display of imagery in the map confirms that all images and necessary metadata have been provided to initiate the workspace. A **Reality Mapping** tab is added to the ribbon.

Tutorial 7-2: Perform a block adjustment

After you create the Reality mapping workspace, the next step is to perform a block adjustment using the tools in the **Adjust** and **Refine** groups. The block adjustment will calculate tie points, which are common points in areas of image overlap. The tie points will then be used to calculate the orientation of each image, known as exterior orientation in photogrammetry.

Calculate the tie points and review the adjustment

1. Click the **Reality Mapping** tab. In the **Adjust** group, click **Adjust**.

    The **Adjust** dialog box appears.

2. Accept the default values for all the settings and click **Run**.

    Note: Depending on your computer, processing may take time.

3. After the adjustment is complete, in the **Contents** pane, turn on the **Tie Points** layer to view the distribution of generated tie points on the map.

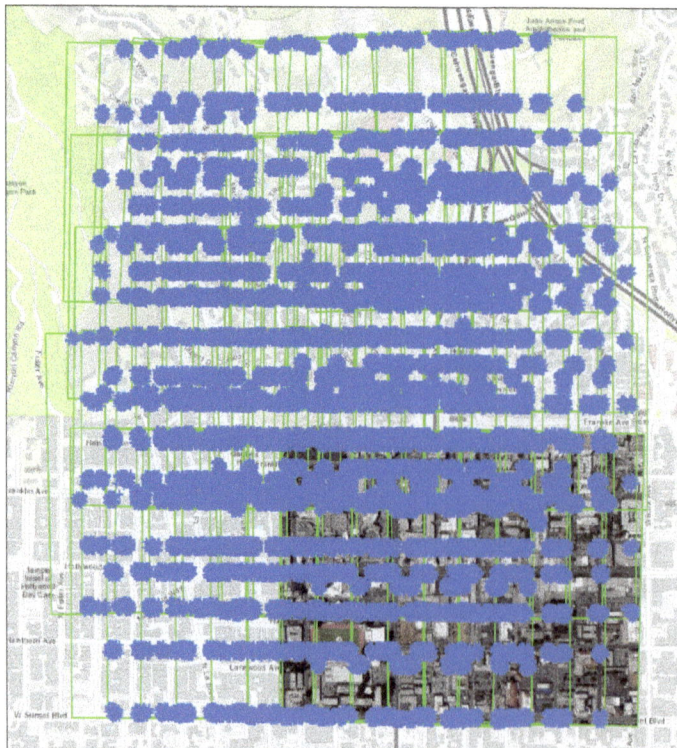

You can view tie point residuals and accuracy reporting in the logs file.

4. Turn off the **Tie Points** layer.

5. In the **Logs**, view the results for the **Computing Block Adjustment**.

```
Computing Block Adjustment...
Start Time: Sunday, August 17, 2025 10:43:39 AM
AverageGSD(m) = 0.0680
MeanReprojectionError(pixel) = 0.396
NumAdjustedImage = 64
NumGPS = 64
NumRemovedImage = 0
NumSolutionPoint = 47361
NumTiePoint = 149965
RMSEGPS = ["0.079","0.070","0.100"]
```

6. On the **Reality Mapping** tab, in the **Review** group, click **Adjustment Report** to generate adjustment statistics.

esri

Adjustment Report

Project Name: RealityMappingAerial_2D

Adjustment Summary

Project Name	RealityMappingAerial_2D
Report Time	2025-8-17 11:10:56
Number of Input Images	64
Number of Adjusted Images	64
Spatial Reference	NAD_1983_2011_StatePlane_California_V_FIPS_0405/VCS:NAVD_1988
Number of Tie Points	149965
Number of Solution Points	47361
Mean Reprojection Error / Sigma Naught (Pixel)	0.37 / 0.51
Ground Resolution (m)	0.064
Adjustment Type	Frame

Quality Checks

Tie Point Mean Reprojection RMSE (pixel)	0.37	✓

The adjustment report provides a record of the adjustment and overall quality measures of the process.

7. Close the report when you are done reviewing it.

Tutorial 7-3: Generate Reality mapping products

Once the block adjustment is complete, imagery products can be generated using the tools in the **Product** group on the **Reality Mapping** tab. The types of products that can be generated depend on various factors, including the sensor, data flight configuration, and scenario type. The flight configuration of this dataset is nadir, which is ideal for 2D products, such as DSM, true ortho, and DSM mesh.

The **Multiple Products** wizard guides you through the workflow to create multiple Reality mapping 2D products in a single process. All generated products are stored in product folders of the same name under the **Reality Mapping** category of the **Catalog** pane.

To generate products using the **Multiple Products** wizard, complete the following steps.

Create multiple 2D products

1. On the **Reality Mapping** tab, in the **Product** group, click **Multiple Products**.

    The **Reality Mapping Products Wizard** appears.

2. In the **Product Generation Settings** step, uncheck the box for **3D**.

    In this tutorial, only 2D products are generated.

3. Under the **2D** products category, uncheck **Digital Terrain Model (DTM)**.

4. Click the **Shared Advanced Settings** button.

    The **Advanced Products Settings** dialog box appears.

5. Change the **Forward Overlap (%)** and **Sideward Overlap (%)** values to 80 and 60, respectively.

6. For **Product Boundary**, click the **Browse** button. Browse to **C:\Top20Imagery \Top20Imagery_07** and click **2D_Prod_Bdry.shp**. Click **OK**.

7. For **Processing Folder**, click the **Browse** button, and navigate to a location on a disk that has a minimum available storage of 11 GB or more available storage.

    > **Note:** The processing folder stores temporary files generated during Reality processing. It is recommended that the processing folder be located on a fast drive with large available storage space.

8. Accept all the other default values and click **OK**.

   The **Advanced Products Settings** window closes, and you return to the **Product Generation Settings** step in the wizard.

9. Click **Next**.

10. In the **DSM Settings** step, click **Output Type** and then click **Mosaic**. Confirm or apply the following settings:
    - **Format**: Cloud Raster Format
    - **Compression**: None
    - **Resampling**: Bilinear

11. Click **Next**.

12. In the **True Ortho Settings** step, set the **Output Type** to **Mosaic** and confirm or apply the following settings:
    - **Format**: Cloud Raster Format
    - **Compression**: None
    - **Resampling**: Bilinear

13. Click **Next**.

14. In the **DSM Mesh Settings** step, under **Formats**, confirm that the box for **SLPK** is checked.

15. Click **Finish** to start the product generation process.

    > **Note:** Product generation may take a few hours, depending on the configuration and resources of your machine.

    Once product generation is complete, the **DSM** and **True Ortho** products are added to the **Contents** pane and 2D map view.

Review 2D products

1. In the **Contents** pane, turn off the **Reconstruction Status** and **Image Collection** layers.

2. Click the **True Ortho** layer and explore the layer.

   The **True Ortho** layer does not contain any building lean, and the tops of buildings, towers, and other elevated features are over their base on the ground.

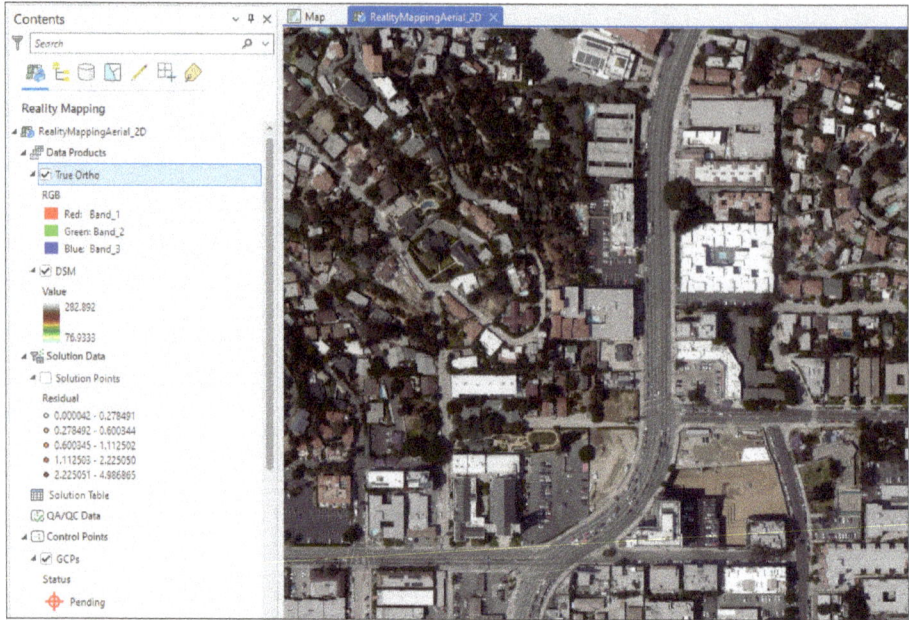

3. Turn off the **True Ortho** layer, click the **DSM** layer, and explore it in the map.

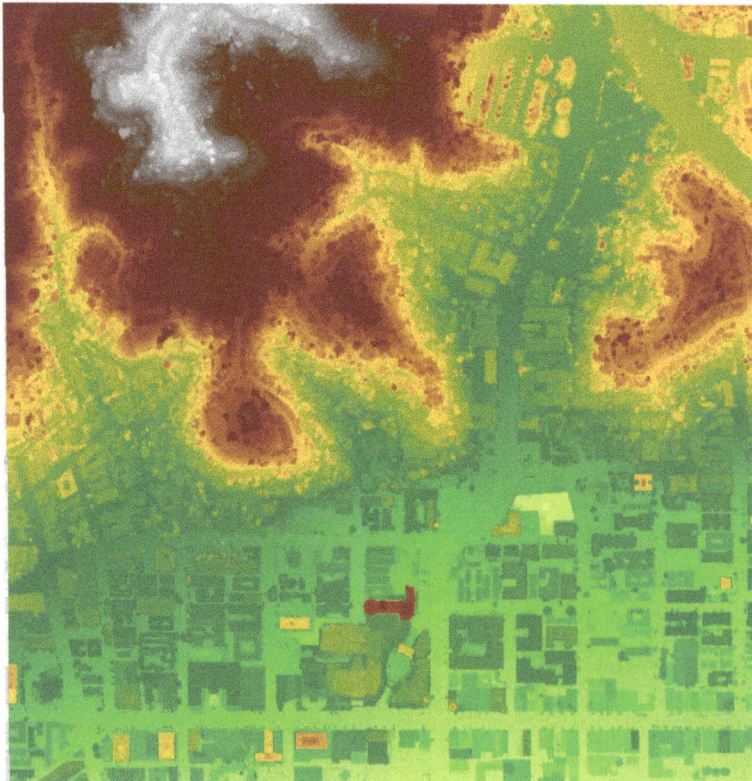

4. On the ribbon, click the **Raster Layer** tab. In the **Rendering** group, click **Symbology**.

   The **Primary symbology** is set to **Stretch**, and the **Color scheme** is set to **Multipart Color Scheme** for visual impact.

5. In the **Catalog** pane, expand **Reality Mapping** > **RealityMappingAerial_2D** > **Products** > **Meshes**.

   The **DSM_Mesh.slpk** product is listed.

6. Right-click the **DSM_Mesh.slpk** file and click **Add to New** > **Local Scene**.

   The **DSM_Mesh.slpk** file is added to a 3D Scene map, with a contextual **Integrated Mesh Layer** tab.

7. Use the 3D navigation tool to tip, tilt, zoom, and roam the 2D mesh layer.

   Note: The 2D mesh is created by draping the orthorectified imagery on the DSM to enable perspective viewing. This is different from a 3D mesh, which is a detailed, geospatially accurate 3D model of a project area or feature with densely and accurately reconstructed ground and facades.

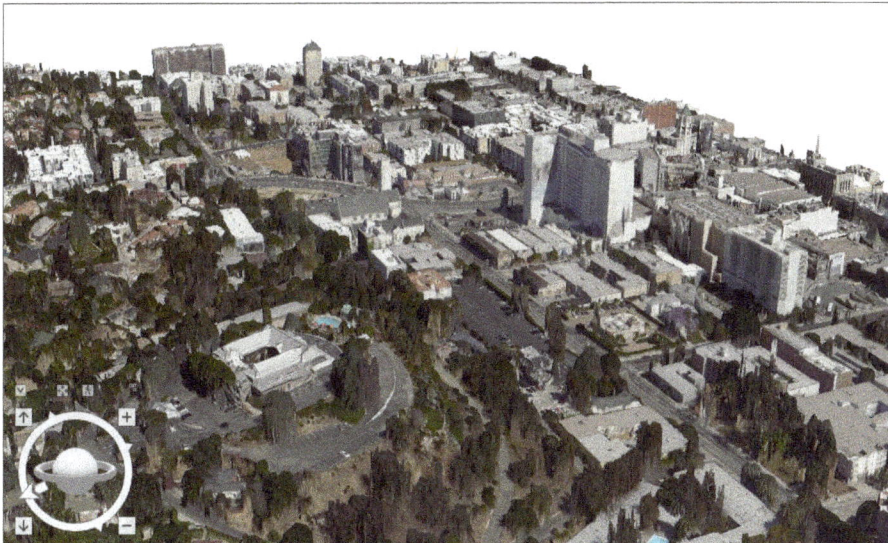

Summary

In this chapter, you learned how to set up a Reality mapping project by creating a Reality mapping workspace and associated image collection. Then you performed the block adjustment and reviewed the Adjustment Report. Lastly, you created three 2D products—a DSM, True Ortho, and DSM mesh—and reviewed them in the map, the Contents pane, and the Catalog pane.

The skills and techniques included in this tutorial can be used to create, process, and manage collections of aerial, drone, and satellite imagery to create Reality mapping products. These products support a wide variety of applications, including GIS land base creation and update, GIS modeling, suitability analysis, and more.

Information at your fingertips

The following links provide background and important information about Reality mapping in ArcGIS Pro.

- links.esri.com/ProRealityMapping_FAQ
- links.esri.com/What_Is_Photogrammetry
- links.esri.com/RealityMapping_in_Pro

PART 2

Understanding and exploring your data

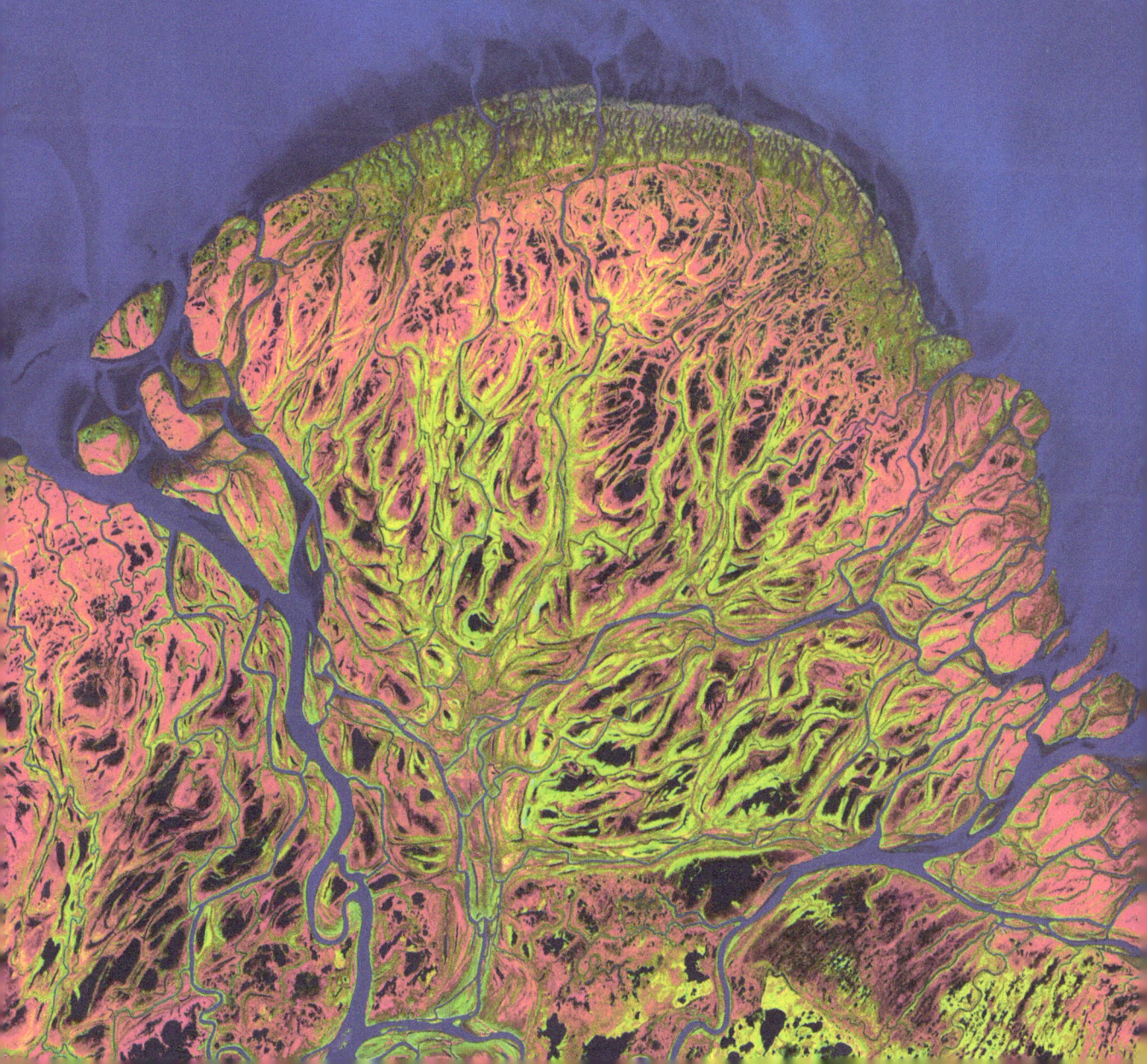

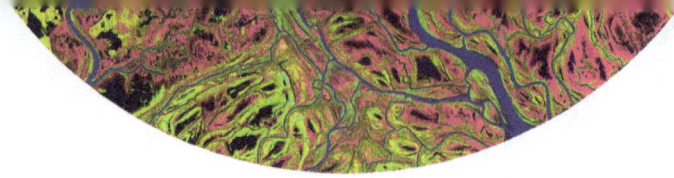

CHAPTER 8
Performing advanced visual analysis

Objectives

- Produce cloud-free imagery using the Pixel Editor.
- Learn to enhance a DEM for visual analysis.
- Work with oblique imagery in Perspective mode.

Introduction

Visual analysis, often referred to as image interpretation, is a key form of image analysis that provides essential context for applications, such as situation awareness and decision support. The primary goal is not to make imagery aesthetically pleasing per se, but to enhance its visual information content and communicative impact. ArcGIS Pro offers a range of tools and capabilities designed specifically to support this kind of visual image information enhancement.

This chapter includes three tutorials, each focused on a different aspect of visual analysis:

1. The first tutorial demonstrates how to use the **Pixel Editor** to remove cloud cover from imagery and replace it with cloud-free data.
2. The second tutorial explores how to enhance digital elevation model, or DEM, data using tools in the **Symbology** pane, raster functions, and rendering effects.

3. The third tutorial guides you through working with oblique imagery to support visual analysis and information extraction.

Tutorial 8-1: Use the Pixel Editor to replace clouds in imagery

Clouds and thick haze obscure ground features important for inventory and management of assets, decision support, and other applications relying on imagery and generally affect imagery collected from satellite platforms. Clouds, haze, and resulting shadows can be replaced with unaffected imagery collected at a different time using the **Pixel Editor**.

In this tutorial, you will replace clouds in Landsat 8 imagery with imagery from a cloud-free Landsat 8 image using the **Pixel Editor**. Follow the steps in this tutorial to replace cloudy regions in your imagery. There are three primary tasks:

- Start a **Pixel Editor** session.
- Replace a cloudy area with clear imagery.
- Save your edits.

Download the tutorial data and set up the project

1. Go to links.esri.com/Imagery20Data and download the data for chapter 8.

2. Unzip the folder to **C:\Top20Imagery**.

   Note: In the second chapter, you created a folder named **Top20Imagery** on your C: drive. If you haven't done that, create that folder now. Now and in subsequent chapters, you will download and unzip the data for each chapter to this folder.

3. In the **Top20Imagery_08** folder, double-click **Top20Imagery_08.aprx** to open the ArcGIS Pro project for this chapter.

   The project contains three folders that contain data for this chapter:

- **PixelEditor**: Contains two Landsat 8 subscenes of the San Francisco Bay Area, each comprising four bands in 8-bit data format.
- **DEM_Rendering**: Contains a NASA Shuttle Radar Topography Mission (SRTM) 16-bit DEM of the San Francisco Bay Area.
- **PerspectiveViews**: Contains a WorldView-3 image of New Orleans, Louisiana, in NITF format.

Replace a cloudy area with clear imagery

1. In the **Catalog** pane, expand **Folders > Top20Imagery_08 > PixelEditor** and add the **CloudFree_8bit_4B.tif** Landsat 8 image to the map.

2. In the **Build Pyramids and Calculate Statistics** dialog box, ensure that boxes under **Pyramids** and **Statistics** are checked. Click **OK**.

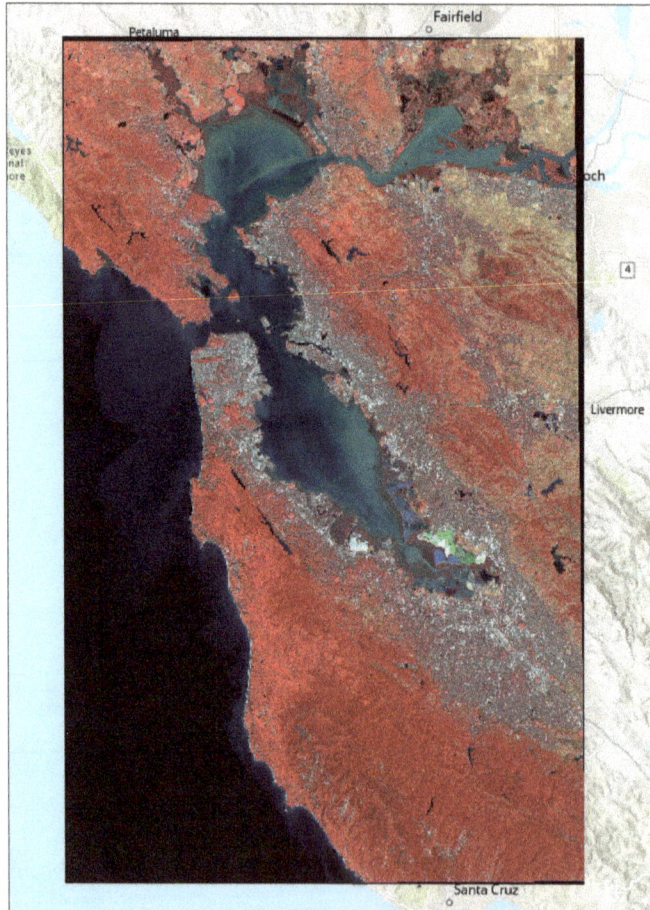

3. Repeat the first two steps for the **Cloudy_8bit_4B.tif** image.

4. On the ribbon, click the **Imagery** tab. In the **Tools** group, click **Pixel Editor**.

   The **Pixel Editor** tab appears on the ribbon, with the necessary tools now available.

5. In the **Contents** pane, turn off visibility for the **CloudFree** image. In the **Cloudy** image, zoom in to a cloudy area.

6. On the **Pixel Edito**r tab, in the **Region** group, click **Mode** and then click **New**.

7. In the **Region** group, click Region and then click Polygon.

8. Draw a polygon around the cloudy area. To complete the polygon, double-click the last point.

   Tip: It is recommended that you digitize along image discontinuities or linear features, such as streets, to help facilitate a seamless replacement.

9. In the **Capture** group, under **Source Layer**, select the layer without clouds (**CloudFree_8bit_4B**).

10. Click **Capture** and then click **Replace**.

   The cloudy area is replaced with imagery from the cloud-free image.

11. Double-click anywhere on the map display to accept the update.

On your own

Remove clouds in other areas of the image by repeating the steps in this section.

Save your edits

To save your edits without affecting the original image, you will save a new file.

1. In the **Save** group, click the **Save as New** button.

2. In the **Export Raster** pane, name the new image Cloudy_Edited.tif and save it to your current workspace. Click **Export**.

> **Tip:** To overwrite your cloudy image with the new cloud-free edits, you would click the **Save** button.

3. In the **Close** group, click **Close Pixel Editor**.

   The **Pixel Editor** closes, and the tab is removed from the main ribbon.

4. If your edited image was added to the **Contents** pane, remove it and leave the original Landsat images. You will use them in the following tutorial.

Tutorial 8-2: Use advanced rendering techniques to visualize DEM data

DEM raster data is often used in conjunction with imagery, providing important 3D information for visual analysis. Follow the steps here to enhance your DEM data.

Assign colors to elevation intervals

1. In the **Catalog** pane, expand **Folders > Top20Imagery_08 > DEM_Rendering** and add the **SRTM30_SanFran_clip.tif** DEM to the map. Accept the settings to build pyramids and calculate statistics.

   The DEM is loaded in the map and listed in the **Contents** pane.

2. In the **Contents** pane, click the **SRTM30_SanFran_clip** layer to enable it and then click the **Raster Layer** tab. In the **Rendering** group, click **Symbology**.

   The **Symbology** pane appears.

3. In the **Symbology** pane, under **Primary symbology**, click **Classify**.

4. Make sure that **Method** is set to **Natural Breaks (Jenks)**.

5. Set the **Color scheme** to **Condition Number**, which ranges from green to red.

   > **Tip:** To see the names of the color schemes, expand the drop-down list and check the box next to **Show names**.

   The DEM is rendered from green (low elevation areas) to red (high elevation areas).

6. Set **Classes** to 10.

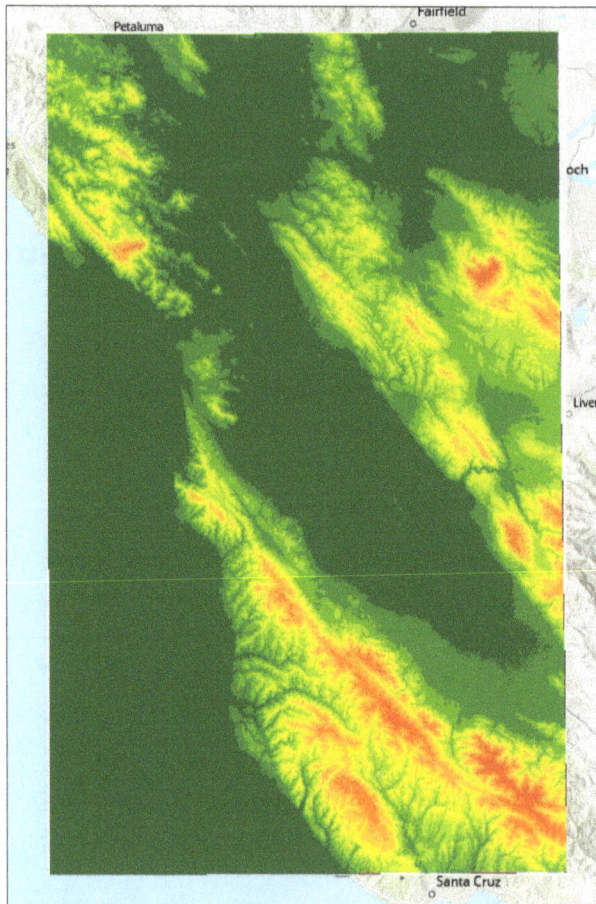

The level of detail visible in the rendered DEM has increased significantly.

7. In the **Classes** table, examine the breakpoints for the color ramp. The lowest value in the DEM dataset is −66.999 meters, whereas the highest value is 1,160 meters.

Next, you will manually adjust the breakpoints and colors.

8. In the first row of the table, in the **Upper value** field, double-click the existing value. Type 0 and press **Enter**.

9. Click the green color chip and, from the color palette, select a deep blue.

This changes the color for mean sea level, and below, to blue.

10. Double-click the second interval. In the **Upper value** field, type 30 and press
    Enter.

    Low-lying areas near sea level are now dark green.

11. Double-click the third interval. In the **Upper value** field, type 100 and press
    Enter.

12. Double-click the fourth interval. In the **Upper value** field, type 300 and press
    Enter.

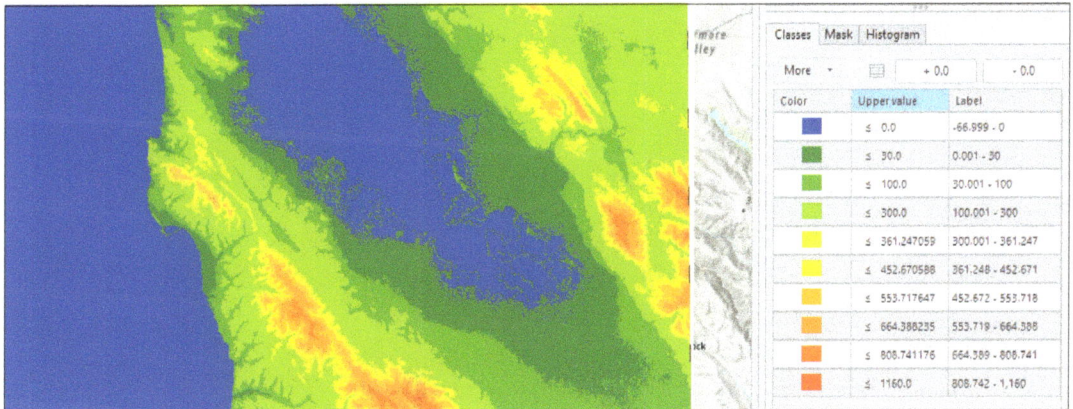

Additional classes and colors could be manually configured if you wanted to
finalize your visualization using this method.

Visualize elevation using shaded relief

1. At the top of the **Symbology** pane, change the **Primary symbology** to **Shaded
   Relief**.

   The DEM is now rendered as a hill-shaded raster. The default setting for
   Hillshade Type is **Traditional**, which simulates a single light source (the sun)
   depending on the specified azimuth (direction) and altitude (degrees above
   horizon).

2. Change the **Hillshade Type** to **Multidirectional**.

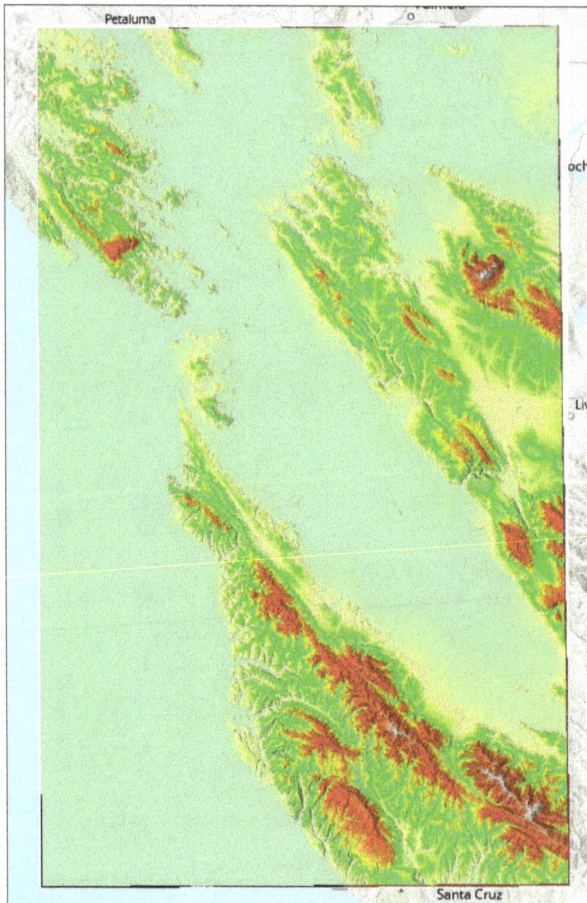

Multidirectional hillshade combines light from multiple sources to represent an enhanced visualization of the terrain.

Enhance an image with hillshade

1. At the top of the **Symbology** pane, change the **Primary symbology** to **Stretch**.

   The DEM is rendered in grayscale.

2. On the **Imagery** tab, in the **Analysis** group, click **Raster Functions**.

   The **Raster Functions** pane appears.

3. In the **Raster Functions** pane search bar, type hillshade. Click the resulting **Hillshade** raster function.

4. In the **Hillshade Properties** tool, apply the following settings:
   - **Raster**: SRTM30_SanFran_clip.tif
   - **Hillshade Type**: Traditional
   - **Scaling**: Adjusted
   - **Z Factor**: 2

5. Click **Create new layer**.

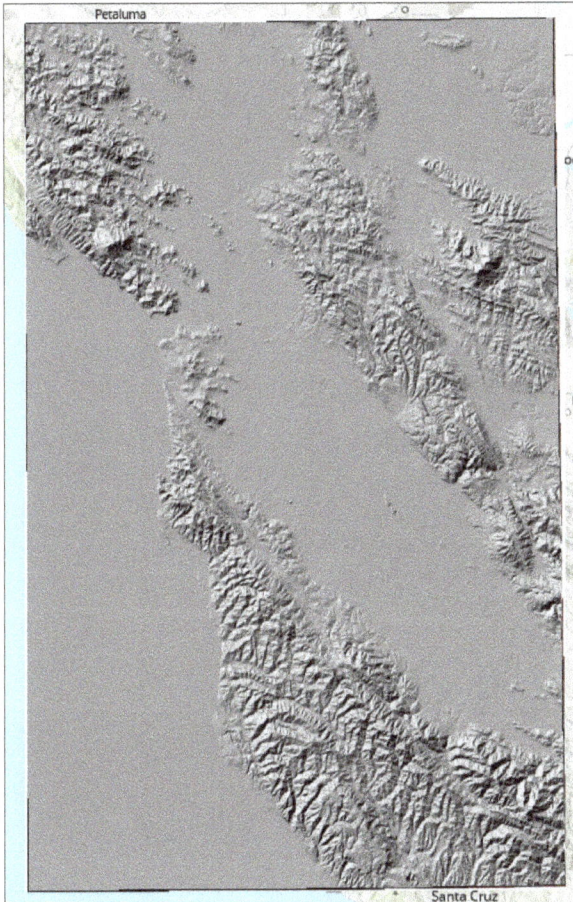

The hillshade DEM is rendered in the map and listed in the **Contents** pane.

6. In the **Contents** pane, turn off all layers except for the **Hillshade_SRTM30_ SanFran_clip.tif** layer and the **CloudFree_8bit_4B.tif** layer.

7. Click the **Hillshade_SRTM30_SanFran_clip.tif** layer to enable it and then click the **Raster Layer** tab.

8. In the **Effects** group, click **Transparency** to expose the transparency slider.

9. Click the slider handle and slide it to adjust the transparency percentage value.

   The transparency of the **Hillshade_SRTM30_SanFran_clip.tif** layer is adjusted to expose the **CloudFree_8bit_4B.tif** layer. This results in a visual combination of the two layers.

10. For good results, adjust the transparency value to around 50–60%.

11. In the **Rendering** group, click **Stretch type** and then click **Esri**. If a warning message pops up asking to compute statistics, click **Yes**.

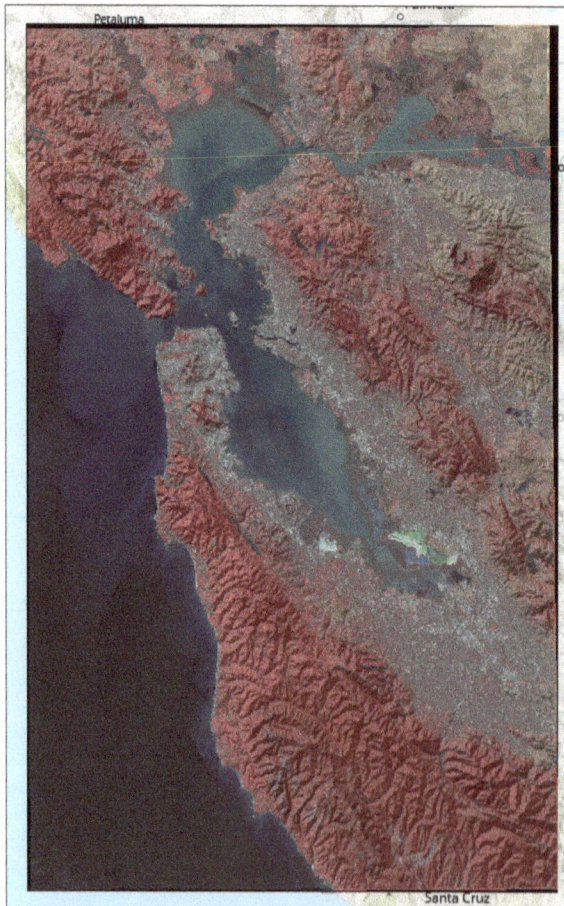

12. Try different stretch types such as **Percent Clip**, **Standard Deviation**, and others.

13. Save your project.

Tutorial 8-3: Work with oblique imagery in Perspective mode

Most high-resolution satellite imagery is collected using agile space platforms that look off-nadir, meaning they collect oblique imagery. Other imaging systems, such as drones, also collect oblique imagery, which is valuable for imaging building facades and other infrastructure, and providing perspective views that are useful for reconnaissance, surveillance, and situation awareness for a variety of applications, such as emergency management and response.

Because oblique imagery is collected from a variety of viewing angles, displaying oblique imagery in map space with north oriented up causes buildings and other ground features to appear to lean at a variety of disorienting angles, making oblique imagery hard to interpret. For example, the following satellite image shows an oblique image displayed in traditional map coordinate space, with north up and the image orthorectified to the basemap.

WorldView-1 image displayed in a map coordinate system, with north at the top of the display. Satellite imagery ©2018 Vantor.

Note how the buildings in the image displayed in map space lean to the northeast, making image interpretation difficult.

The following satellite image shows the oblique image in image coordinate space, undistorted and rotated with buildings displayed vertically in a more natural orientation.

WorldView-1 image displayed in an image coordinate system, with north to the right of the display. Satellite imagery ©2018 Vantor.

ArcGIS Pro provides the capability to display and work with oblique imagery in image coordinate space (ICS), called perspective viewing. This helps you be more effective in addressing image interpretation applications. Working with oblique imagery in image space facilitates seamless transformation between the image coordinate system and map coordinate system to display, interpret, collect, and edit GIS feature data in the oblique imagery.

In this tutorial, you will visualize imagery, capture data, and perform vertical measurements on imagery in **Perspective** mode. The tutorial has the following sections:

1. Prepare your ArcGIS Pro project.
2. Activate image space.
3. Visualize your data in image space.
4. Make vertical measurements on imagery in image space.

Prepare your ArcGIS Pro project

1. On the ribbon, click the **Insert** tab. In the **Project** group, click **New Map**.

2. In the **Catalog** pane, expand **Folders > Top20Imagery_08 > PerspectiveView** and add the **NOLA.NTF** image to the map. When prompted to build pyramids for this image, click **Yes**.

   > **Note:** The **NOLA.NTF** image is a portion of a larger image in NITF format. The technical name of this image is **18NOV15171817-P2AS_R3C3-010246901050_01_ P001.NTF**, which is the name displayed in the **Contents** pane. For simplicity, this image will be referred to as the NOLA image in the tutorial.

Satellite imagery ©2018 Vantor.

The image is displayed in standard cartographic map view, with north up.

3. In the **Contents** pane, click the expander next to the image name to display image details.

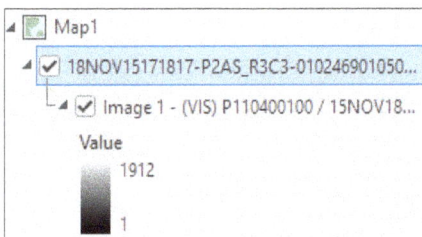

Activate image space

1. At the top of the **Contents** pane, click the **List By Perspective Imagery** button.

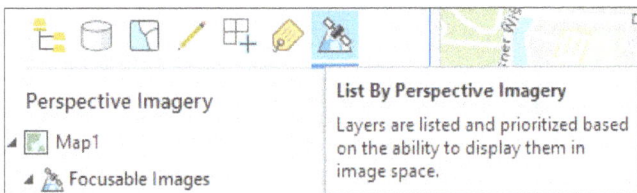

The images are sorted based on focusable images. Focusable images are those eligible for display and analysis in image space, determined by required metadata associated with the imagery. The focusable images are listed first under the heading **Focusable Images**, followed by **Others** layers listed in the **Contents** pane.

Next, you will view the imagery in image space map view.

2. In the **Contents** pane, select the NOLA (**18NOV15171817-P2AS_R3C3-010246901050_01_P001.NTF**) layer.

3. Next to **Image 1**, click the **Set as focus image** button.

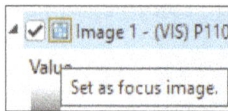

4. On the **Raster Layer** tab, in the **Rotation** group, click **Rotation Type** and then click **Up is Up**.

5. In the **Rendering** group, make sure that **DRA** is not selected.

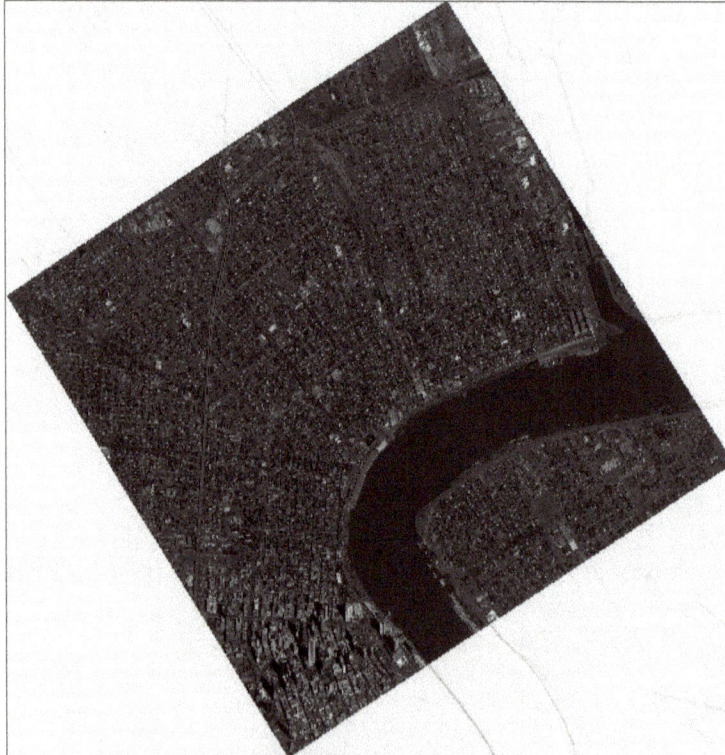

Satellite imagery ©2018 Vantor.

The image in the map rotates so that buildings are oriented vertically. Now you will be able to visualize and work with the imagery from the perspective of the sensor.

The map now has a heads-up display, or HUD, window overlaid, which gives you easy access to commonly required metadata pertaining to the imagery.

6. Zoom in to the downtown area with tall buildings, located in the bottom corner of the image.

7. Move the cursor to the edge of the HUD to change the cursor type to a four-headed arrow, which lets you move and place the HUD anywhere on the map based on your preference.

8. In the top right corner of the HUD, click the arrow to minimize it.

Measure building heights in imagery

1. In the **Contents** pane, make sure the NOLA image is active.

2. On the **Imagery** tab, in the **Mensuration** group, click the list to access the mensuration tools.

3. In the **Mensuration** group, click the **Mensuration Options** button to adjust settings such as measurement units.

The tools available depend on the metadata of the image layer selected in the **Contents** pane. The NOLA image contains satellite platform and sensor orientation information, which enables 3D measurement tools.

4. Click the **Base To Top Height** measurement tool.

5. On the map, click to place a point at the base of any building and then double-click to place a second point at the top of the building.

The **Mensuration Results** pane appears and displays the height measurement and other associated metadata information.

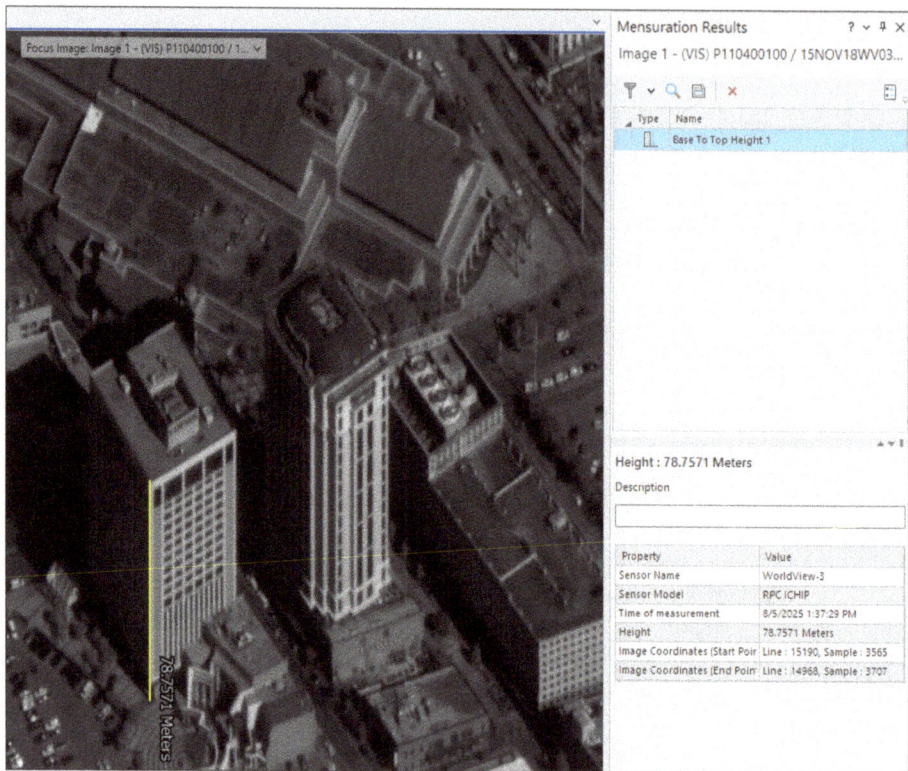

Satellite imagery ©2018 Vantor.

6. Measure a few more buildings using the **Base To Building Height** 3D measurement tool.

    All the measurements are recorded and listed in the **Mensuration Results** pane.

    > **Tip:** If you need to pan to a different area of the image, click the **Map** tab on the ribbon and then click the **Explore** button in the **Navigate** group. Click the **Imagery** tab to access the measurement tools.

7. Select the **Base To Top Shadow Height** 3D measurement tool.

8. In the map, pick two points. Make sure the first point is the base of the building structure and the next is the top of the shadow cast by the building top.

    > **Tip:** Make sure that the shadow you measure is cast on the ground and not on another building or elevated feature.

Satellite imagery ©2018 Vantor.

9. Collect a few 2D measurements, such as road distance and the area of parks.

10. In the **Mensuration Results** pane, click the **Generate Report** button to export the mensuration results to a text file in a specified folder. Name the report Mensuration Report and click **Save**.

> **Note:** Measurements are automatically saved to a feature dataset in the project geodatabase. The measurements are loaded automatically into the **Mensuration Results** pane when a project is loaded.

Summary

In this chapter, you learned how to enhance imagery in a variety of ways for improved visual analysis. You used the Pixel Editor to replace clouds in imagery with cloud-free imagery, enhanced a DEM with a color ramp with specified intervals, and created a hillshade raster used to enhance a multispectral image. Finally, you transformed oblique imagery from map space to image coordinate space and then measured features in the transformed imagery. These techniques improve the visual information content in imagery and raster data and help facilitate more effective and timely situation awareness and decision support.

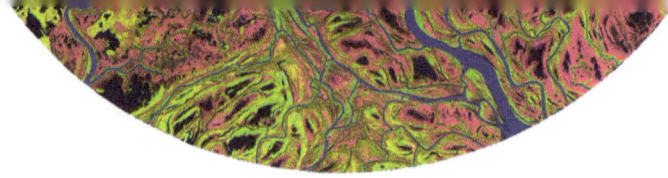

CHAPTER 9
Exploring image charts for analysis

Objectives

- Create spectral profiles.
- Create and examine scatter plots.

Introduction

In this chapter, you'll learn how to use charts for image analysis. Every pixel in an image is a spectral and spatial measurement representing a location in the geography. More than three bands of information often represent a sample of the spectral signature, or spectral profile of features being displayed. Because of these unique spectral characteristics of imagery, you can see and examine much more than just the visual representation of the image on a map. Viewing this information as special types of charts can help make things even clearer, especially when comparing the fine details related to two or more features of interest or examining patterns through time.

First, you'll create spectral profiles to help identify and discriminate various features on the surface. There are several ways to represent this spectral information using spectral profiles. You'll learn about two of them in this chapter.

Tutorial 9-1: Create a spectral profile

Spectral profile charts allow you to select regions or ground features on the image and review the spectral information for all the bands in the image. Natural and human-made materials often have unique spectral signatures that can be used to identify them quantitatively. Spectral signatures such as this provide a "spectral fingerprint" for features of interest in an area or region of interest. For instance, if you're comparing two agricultural fields to determine whether they contain the same crop, you can select an area from each field and display the summary of pixel values from each band for each field in the spectral profile chart.

In this first tutorial, you'll use spectral profiles to examine features in a region in Germany.

Download the tutorial data and set up the project

1. Go to links.esri.com/Imagery20Data and download the data for chapter 9.

2. Unzip the folder to **C:\Top20Imagery**.

> **Note:** In the second chapter, you created a folder named **Top20Imagery** on your C: drive. If you haven't done that, create that folder now. Now and in subsequent chapters, you will download and unzip the data for each chapter to this folder.

3. Inside the **Top20Imagery_09** folder, double-click **Top20Imagery_09.aprx** to open the ArcGIS Pro project for this chapter.

Prepare the image for collecting spectral profiles

You will familiarize yourself with the layers that have already been added to the project.

This project contains two maps. The first map, **Landsat 8 OLI (Germany)**, contains a subset of a Landsat 8 image collected on April 3, 2025, of the Rhein-Neckar Kreis district of northwest Baden-Württemberg, Germany. This map also contains a feature layer called **SpectralAreas**. You will use this layer to help collect spectral profiles. The second map, **Landsat 5 TM (Germany)**, contains an older Landsat 5 image of the same area collected on July 6, 2001. You will use this image in the tutorial when creating scatter plots, also known as scatterplots, of analytical results.

Before you create spectral profiles, however, you will need to calibrate the Landsat 8 image to reflectance values.

> **Note:** See chapter 5 for detailed information on atmospheric calibration and converting images to reflectance values.

1. Make sure the current map is **Landsat 8 OLI (Germany)**. On the ribbon, click the **Imagery** tab. In the **Analysis** group, click **Raster Functions**.

2. At the top of the **Raster Functions** pane, click the **Find Raster Functions** field and type Apparent Reflectance.

3. Click the tool and apply the following settings:
    - **Raster**: Multispectral_LC08_L1TP_195026_20250403_20250411_02_T1_MTL
    - Check the box for **Albedo**.

Raster

Multispectral_LC08_L1TP_195026_20250403_202 ⌄	📁

Reflectance Gain and Bias Values per Band (Watts per square meter per micron)

	Reflectance Gain	Reflectance Bias
1	2E-05	-0.1
2	2E-05	-0.1
3	2E-05	-0.1
4	2E-05	-0.1
5	2E-05	-0.1
6	2E-05	-0.1
7	2E-05	-0.1
8	2E-05	-0.1
*		

Sun Elevation (degrees)

43.90210873

☑ Albedo

Scale Factor

50000

Offset

5000

4. Click **Create new layer**.

Now that your image is using reflectance values, you can create a spectral profile.

Create spectral profiles of a pixel

The first type of spectral profile you will create is a mean line plot. A chart of this type creates a line connecting the mean values within a set of pixels per band. This means that the pixel values of the image bands that intersect an area of interest will be plotted. You can define an area of interest several ways, such as a point, a line, a polygon, or even from a feature layer already in your map. When you use a point as your area of interest, you are creating a spectral profile for a unique pixel. Your first spectral profile will use a point and represent a single pixel.

1. On the **Map** tab, in the **Navigate** group, click **Bookmarks** and select **Spectral Profiles Fields**.

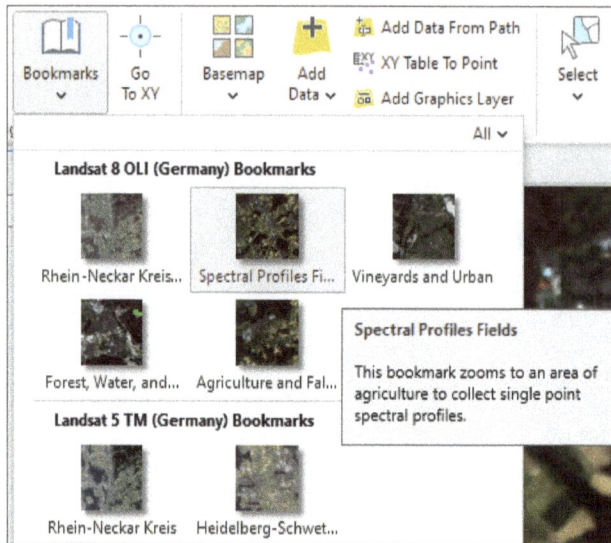

You will select a few of these agricultural fields as a point (or single pixel) spectral profile.

2. In the **Contents** pane, right-click **Apparent Reflectance_ Multispectral_ LC08_L1TP_195026_20250403_20250411_02_T1_MTL**. Hover over **Create Chart** and click **Spectral Profile**.

The **Chart Properties** pane appears as well as a blank chart below your map that fills in once you create your profiles. The image highlights the three areas of agricultural fields that you will use to create spectral profiles. Use this image as your reference.

3. In the **Chart Properties** pane, under **Define an area of interest**, click the **Point** button.

4. In the map, move your cursor to the westernmost field in the graphic in step 2 and click to select a point in the brown/fallow agricultural field.

Your cursor will become a crosshair with a circle for the point.

After you select a point in the field, the chart profile is created automatically. In addition to the chart view, the chart is also added to the **Contents** pane underneath the **Apparent Reflectance** results as part of **Charts** and is named **Spectral Profile 1**.

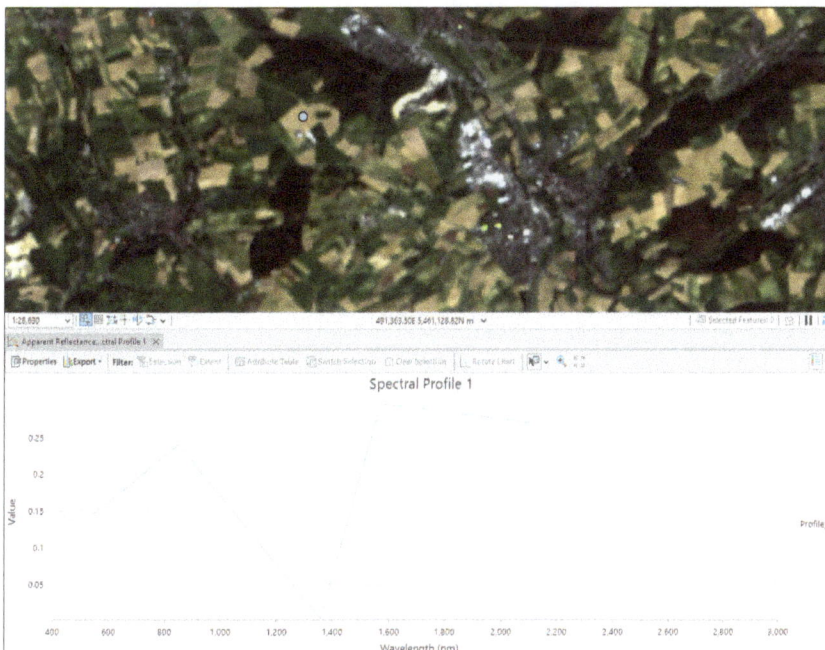

Tip: Your spectral profile may appear slightly different from this example depending on which pixel you selected. Also, the color of the profile may be different. ArcGIS Pro assigns the colors randomly.

The axes labels are assigned automatically. The x-axis in wavelength (nm) is derived from the raster product metadata, and the y-axis is set as reflectance values computed from the apparent reflectance function.

Refine your chart and view spectral profiles

From the **Axes**, **Format**, and **General** tabs, you can control how the chart appears. You can also set the label and the color to help differentiate this profile from others you collect.

1. In the **Chart Properties** pane, in the **Spectral Profiles** table, under the **Label** column, click **Profile_1** and type Fallow Agriculture.

2. Under **Symbol**, click the color patch and then click **Electron Gold** (column 4, row 3).

   After you make these two changes, the label in the chart changes as does the color of the line segment.

3. Using skills you just learned, collect two more spectral profiles using the boxed areas in the reference image as guides.

4. Label the spectral profile for the northernmost field North Field. Label the spectral profile for the southernmost field South Field.

5. Set the color for the two new spectral profiles as follows:
   - **North Field**: Tarragon Green (column 6, row 5)
   - **South Field**: Macaw Green (column 6, row 4)

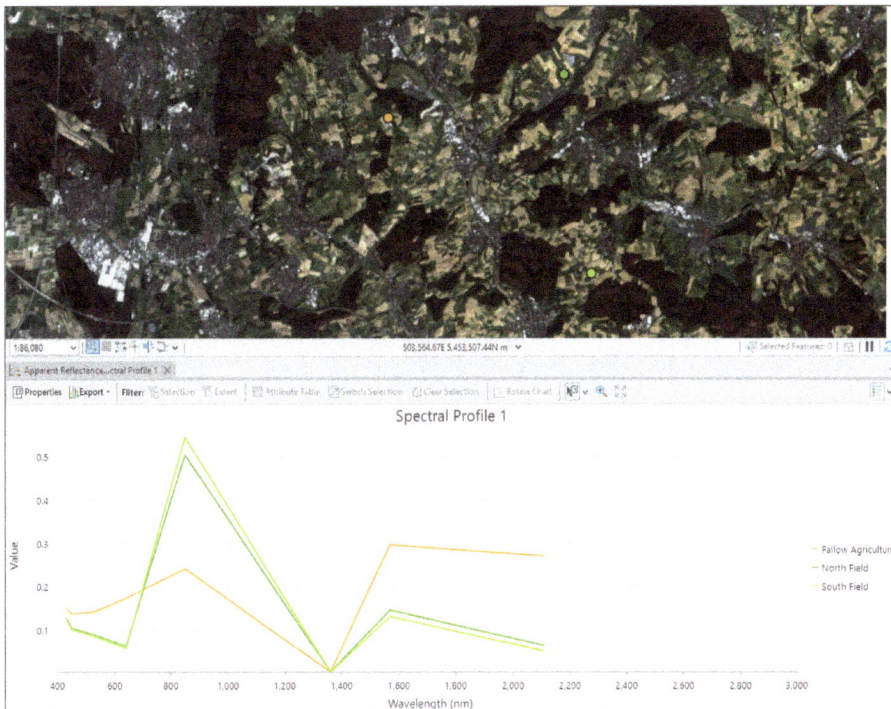

Each sharp point in these curves represents the center of one of the eight spectral bands of the Landsat 8 OLI sensor. Although it's easy to see the difference between the **Fallow Agriculture** field profile and the other two profiles, the other two profiles (**North Field** and **South Field**) are similar. However, there are a few differences—points of separation—in the two Shortwave Infrared bands (band 6 and band 7) around 1,580 nm and 2,100 nm. These differences may aid in helping to discriminate crop types or other material characteristics from one another.

Delete spectral profiles

Now you'll delete these three spectral profiles to collect other profiles in the area.

1. In the **Chart Properties** pane, in the **Spectral Profiles** table, in the **Enabled** column, click the **Fallow Agriculture** row to select it.

2. Click the **Delete The Selected Row** button.

3. Repeat the previous step to delete the **North Field** and **South Field** profiles.

    As you delete each profile, the profile (line) segment in the chart disappears. After all the profiles are deleted, the chart is once again blank.

    Next, you'll collect and analyze spectral profiles collected from a feature layer.

Create spectral profiles from a feature layer

You can collect spectral profiles for areas of interest that appear to be of a similar material type or natural feature. For instance, instead of creating a spectral profile of a single pixel in an agricultural field, you can create a spectral profile that combines the spectral information within a polygon feature class or along a line segment. In this step, you'll use the feature layer **SpectralAreas** to collect spectral profiles and examine their differences and similarities.

1. In the **Contents** pane, select the **SpectralAreas** layer and turn on its visibility.

2. On the **Map** tab, in the **Navigate** group, click **Bookmarks** and select Vineyards and Urban.

    Because you are creating a spectral profile from a set of pixels, you will need to change the way the information is represented.

3. In the **Chart Properties** pane, under **Plot Type**, click Boxes And Mean Lines.

4. Under **Define an area of interest**, click the **Feature Selector** button.

5. Move your cursor onto the map and click the **Vineyard** polygon to select it.

    > **Tip:** Use the legend of the feature layer in the **Contents** pane to help identify which polygon is the vineyard.

    This time the spectral profile looks a bit different from before.

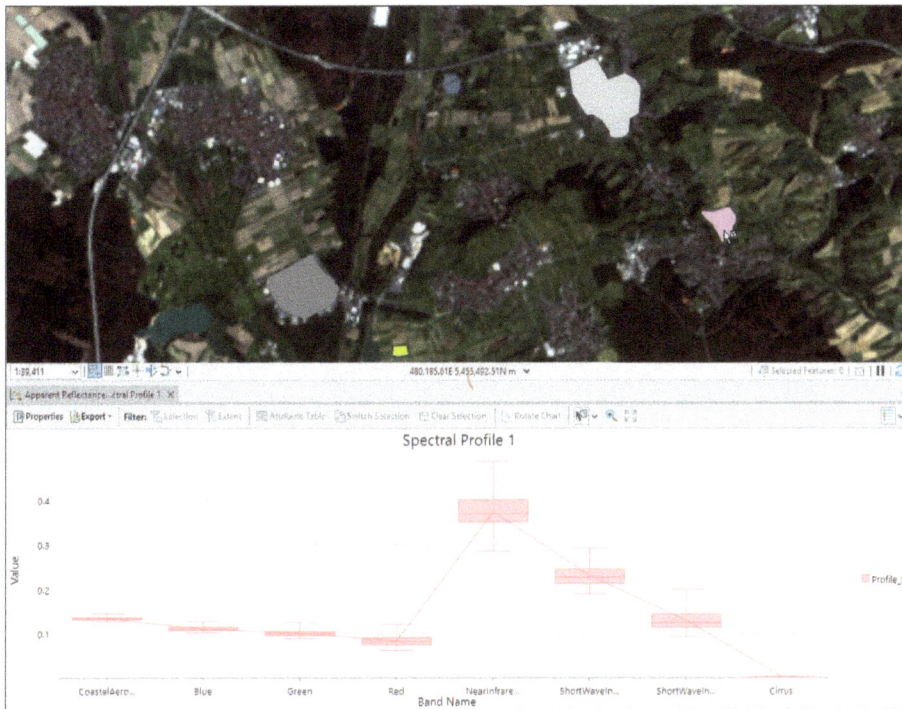

Note: The color and name of your profile may be different from the graphic.

There are two distinct differences between this new chart and the one you previously created. The first, and most obvious, is the representation of the spectral profile using the box and mean line method. This profile plots boxes and a line graph connecting the mean values within the set of pixels making up the entire area of interest—in this case, your selected feature. It combines the mean line like what you created before, but it also combines boxes for each area of interest. The box type allows you to visualize and compare the distribution and central tendency of the set of pixels collected through their quartiles. Quartiles are a way of categorizing numeric values into four equal groups based on five key values: minimum, first quartile, median, third quartile, and maximum. In the chart, you can place your cursor over the box to see these values.

The second difference is more subtle. It is how the x-axis is represented, as a by-product of the **Boxes and Mean Line** method. Unlike the previous spectral profile where the x-axis was the spectral range in nanometers, this profile is by band name. The Cirrus band is now located at the extreme right end of the chart.

The spectral range for the Cirrus band is 1,360–1,380 nm, so in the original chart, it falls between the Near-Infrared band and the two Shortwave Infrared bands.

6. Using skills you previously learned, change the **Label** of the profile to Vineyard and, if desired, the color.

7. Now collect two more spectral profiles in this area for the **Low Density Urban** and **Medium Density Urban** polygons.

In these two urban profiles, notice the range and mean box differences for the two features. ArcGIS Pro also automatically offsets these boxes to help make them easier to see and examine. This is why the band names are used for the x-axis instead of wavelengths.

8. On the **Map** tab, in the **Navigate** group, click **Bookmarks** and select **Forest, Water, and Agriculture**.

9. Using skills previously learned, collect four more spectral profiles in this area for **Agriculture 1**, **Forest**, **Forest (Deciduous)**, and **Water** polygons.

You will now collect spectral profiles for the final two features in your map.

10. On the **Map** tab, in the **Navigate** group, click **Bookmarks** and select Agriculture and Fallow.

11. Collect the final two spectral profiles in this area for the **Agriculture 2** and **Agriculture (Fallow)** polygons.

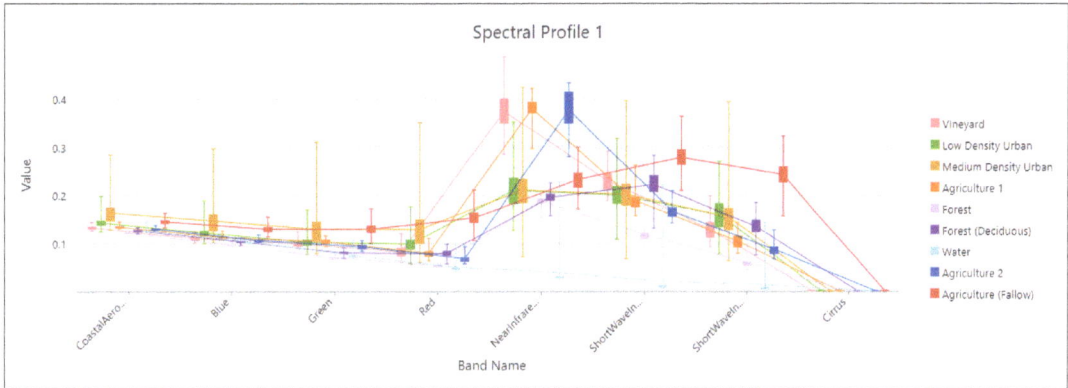

Looking at the spectral profiles like this can be interesting, but it might not be the easiest way to examine and analyze differences and similarities between features. In the next step, you learn how to set a few preferences to examine the spectral profiles of your choice while maintaining the other profiles in your chart.

Examine and compare spectral profiles

Next, you'll examine the spectral profiles of the vineyard area and the three agricultural areas and exclude the other profiles.

1. In the **Chart Properties** pane, click the **General** tab, and under **Chart Title**, type Rhein-Neckar Kreis Spectral Profiles and press **Enter**.

   In addition to the chart title changing in your profile, the name change is also reflected under **Charts** in the **Contents** pane.

   You can float the chart view to make it easier to examine.

2. In the **Spectral Profiles** chart, right-click the tab and click **Float**.

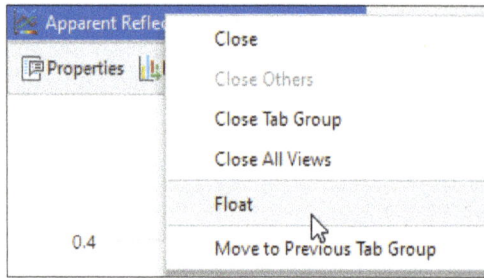

Now you can control the chart the same way as any other window to make it bigger or smaller.

3. Expand the **Spectral Profiles Chart** window but make sure you can still view and interact with the **Chart Properties** pane.

4. In the **Chart Properties** pane, under **Spectral Profiles**, uncheck the following profiles:

- **Low Density Urban**
- **Medium Density Urban**
- **Forest**
- **Forest (Deciduous)**
- **Water**

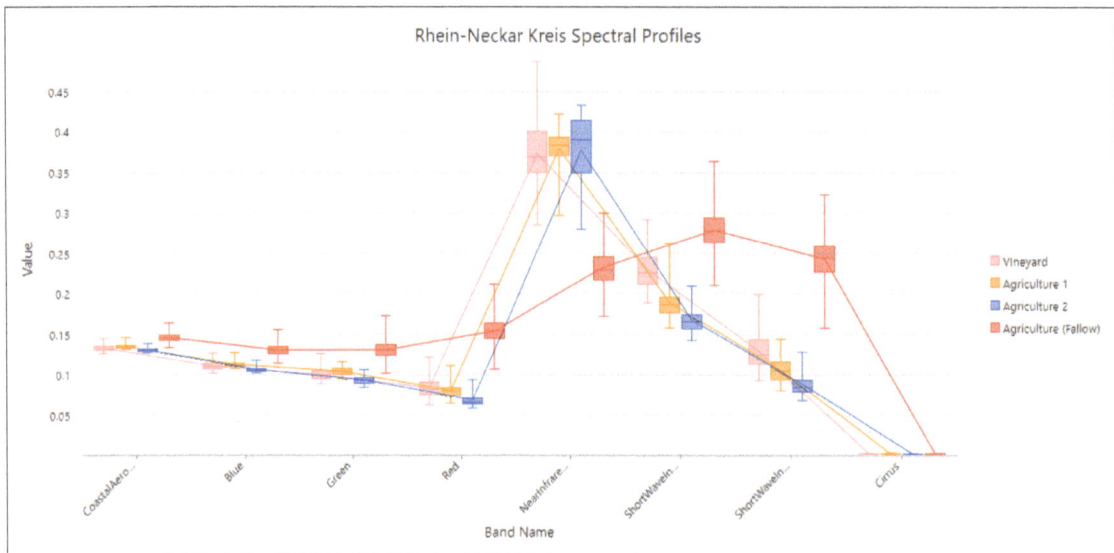

Using techniques such as this, you can compare spectral profiles to help identify unique areas, such as crop types, differences in forested areas, and sometimes even material identification. You can also export the spectral profile as a graphic (.png, .svg, or .jpg format), as a table, or capture it on a clipboard to paste in another application.

5. In the chart, click **Export** and then click **Export As Graphic**.

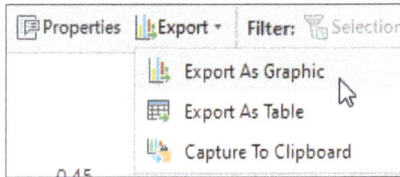

6. In the **Export** window, save your spectral profile to a location and file name of your choice.

7. Close the chart window.

8. In the **Contents** pane, turn off and collapse the **SpectralAreas** layer.

In this tutorial, you learned how to create two types of spectral profiles. You created spectral profiles of individual pixels using the **Mean Line** method. You also created spectral profiles from a feature layer using the **Boxes and Mean Line** method. Each provides unique and useful ways to analyze and understand your imagery.

Tutorial 9-2: Create a scatter plot

In this tutorial, you'll create a scatter plot using the same Landsat 8 image of Germany. Scatter plots portray the spectral information of feature data, allowing you to visualize two bands of information plotted along two axes as a chart. It's a different way to examine and understand an image. This type of chart is used to examine the association between image bands and their relationship to features and materials of interest. To create a scatter plot, the pixel values of one band are displayed along the x-axis, and those of another band are displayed along the y-axis. Features and materials in the image can be identified where the two variables intersect in the distribution.

Create and examine a scatter plot of multispectral image bands

You'll use the same apparent reflectance image of Germany to plot two bands against each other. Often, one of the easiest features to identify in imagery is water. Because of the spectral characteristics of water in the Blue band and Near-Infrared band, you can use a scatter plot to isolate and identify water in your Landsat 8 image.

1. In the **Contents** pane, right-click the **Apparent Reflectance_Multispectral_ LC08_L1TP_195026_20250403_20250411_02_T1_MTL** layer and click **Zoom To Layer**.

2. Right-click the **Apparent Reflectance_Multispectral_LC08_L1TP_195026_ 20250403_20250411_02_T1_MTL** layer again and hover over **Create Chart**. Click **Scatter Plot**.

    As in the previous tutorial, the **Chart Properties** pane appears, as well as a blank chart area. To see a scatter plot, you will need to identify the spectral bands you want plotted against each other.

3. In the **Chart Properties** pane, for **X-Axis Number**, click **Blue**.

4. For **Y-Axis Number**, click **NearInfrared**.

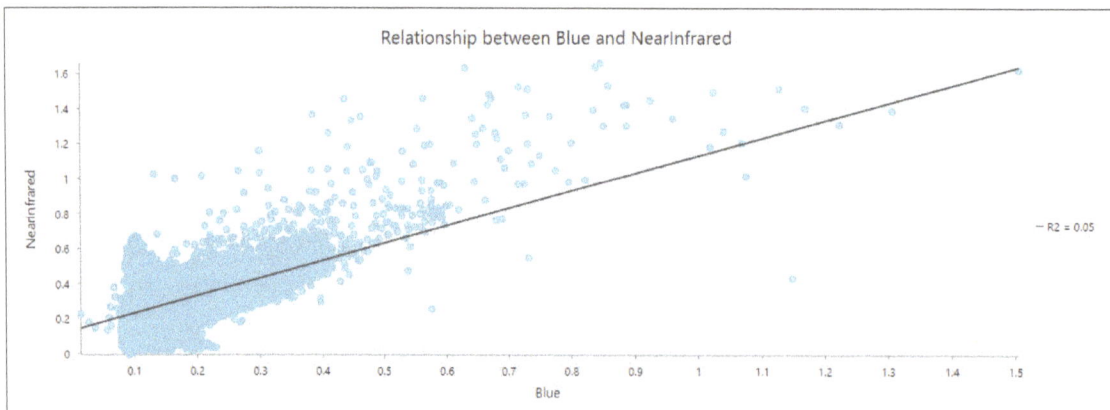

The pixels of the two bands are now plotted against each other. The title and axes labels are added automatically. As part of the statistics of a scatter plot, ArcGIS Pro calculates a regression equation; the associated trend line and R2 value are plotted on the chart. This trend line models the linear relationship between x and y, and the R2 quantifies how well the data fits the model. These elements—the trend line and R2—are relevant only for linear relationships.

You can use tools in the chart to select pixels (areas of pixels) in the scatter plot, and that selection will be highlighted in the map. You can interactively select a region in the distribution and visualize the associated image pixels in the map display. This allows you to visually examine the distribution of pixel values making up features in your scatter plot and visualize the separation between feature classes for a given set of scatter plot variables.

Because water absorbs most of the energy in the Near-Infrared band, most pixels of water in this band have a low reflectance. Inland water features tend to also have low reflectance values in the Blue band. Understanding conventional spectral characteristics of features and how high or low they reflect light helps when identifying features in a scatter plot.

5. On the **Scatter Plot** chart toolbar, click the **Select Interaction Mode** button and then click **Select by polygon**.

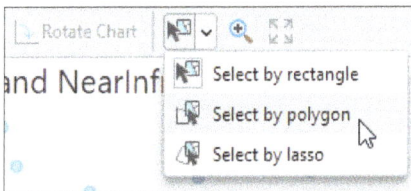

6. On the chart, use your cursor to select the pixels in the bottom left corner of the scatter plot.

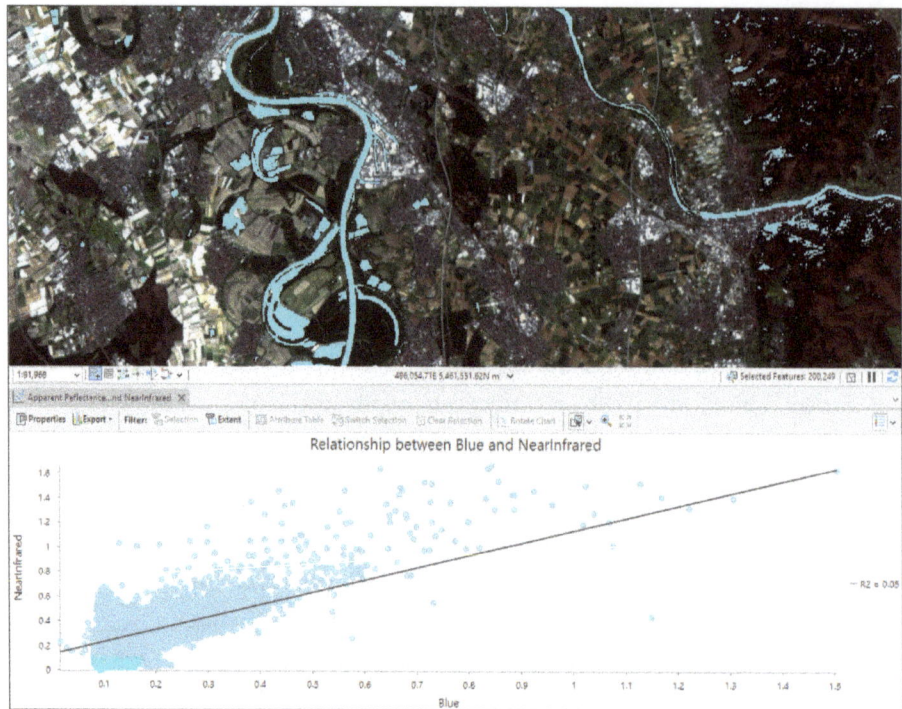

Note: Your graphic may appear slightly different, depending on the pixels and region you selected.

Water features are highlighted on your map. Areas of shadow are probably also highlighted as well. Because shadows also have low reflectance values in both bands, there can be confusion between water and shadows. As in the previous tutorial, you can export this result.

7. On the chart toolbar, click **Export** and then click **Capture to Clipboard**.

8. Open an application of your choice, such as Microsoft Word or Microsoft PowerPoint, and paste the graphic into a blank page or slide.

9. Close the chart window.

In this tutorial, you learned how to use a scatter plot to represent two image bands of information plotted along two axes. You then used knowledge of spectral characteristics to identify and select features in the image based on their pixel values that are plotted in the spectral profile.

Tutorial 9-3: Create a scatter plot from an analysis result

Sometimes imagery can provide a lot of extra information. It can be difficult to know which of the data is the most important. Remote sensing research over the years has helped reduce and transform this dimensionality into smaller components. You can take these analytical results and use them in scatter plots, too, to help identify features.

In this tutorial, you'll use a special type of analysis—a Tasseled Cap (Kauth-Thomas) transformation—in a scatter plot.

Calculate a Tasseled Cap (Kauth-Thomas) transformation

The Tasseled Cap (Kauth-Thomas) transformation is designed to analyze and map vegetation phenomenology and urban development changes detected by various satellite sensor systems. It is known as the Tasseled Cap transformation because of the shape of the graphic distribution of data.

To calculate the Tasseled Cap for this Landsat 5 image, you will use a raster function.

1. Click the **Landsat 5 TM (Germany)** tab to open the map.

2. On the **Imagery** tab, in the **Analysis** group, click **Raster Functions**.

3. In the **Raster Functions** pane, click the **Find Raster Functions** field and type Tasseled Cap.

4. Click the tool. In the settings, for **Raster**, click **Multispectral_LT05_L1TP_ 195026_20010706_20200906_02_T1_MTL**.

5. Click **Create new layer**.

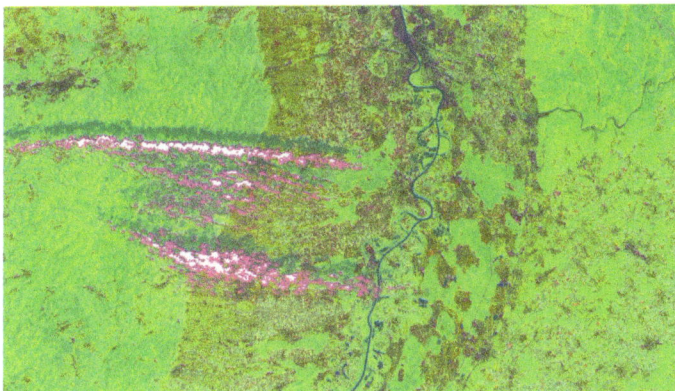

Next, you'll analyze some of the areas in the Rhein-Neckar Kreis district.

6. On the **Map** tab, in the **Navigate** group, click **Bookmarks** and select **Rhein-Neckar Kreis**.

Create a new scatter plot of Tasseled Cap results

Now you'll create a scatter plot of these results. A Tasseled Cap transformation has three primary components: brightness (**Band_1**), greenness (**Band_2**), and wetness (**Band_3**). Your function result displays these three components in a red-green-blue (RGB) combination. Even though these bands do not represent reflectance values, you can still use them to create your scatter plot.

1. In the **Contents** pane, right-click the **Tasseled Cap (Kauth-Thomas)_Multispectral_LT05_L1TP_195026_20010706_20200906_02_T1_MTL** layer. Click **Create Chart > Scatter Plot**.

2. In the **Chart Properties** pane, for **X-Axis Number**, click **Band_1**.

3. For **Y-Axis Number**, click **Band_2**.

4. In the **Chart Properties** pane, on the **General** tab, type the following settings:
    *   **Chart Title**: Tasseled Cap, Landsat 5 TM, July 2001
    *   **X-Axis Title**: Brightness
    *   **Y-Axis Title**: Greenness

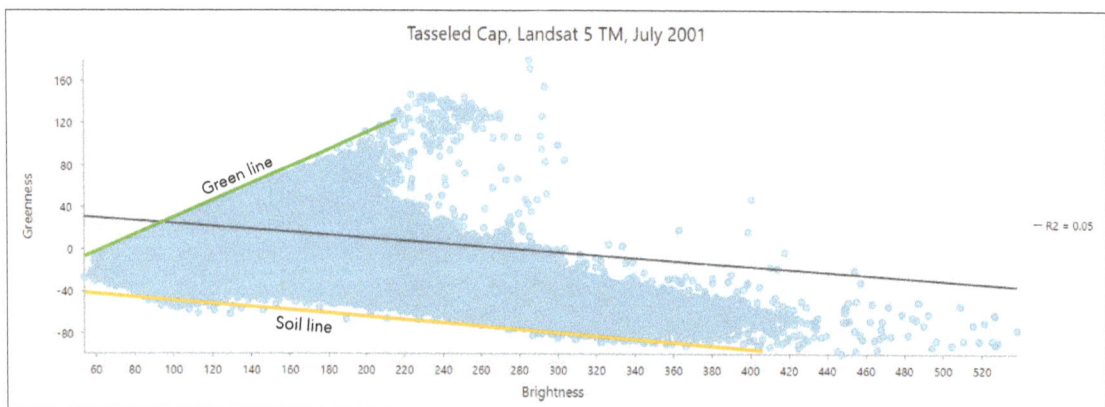

The green line and soil line are for illustration and are not generated by the scatter plot.

The resulting scatter plot displays the Tasseled Cap distribution where different vegetation species are located along the green line, whereas soil types and human-made features are distributed along the yellow line. The Tasseled Cap distribution is also useful for determining the growth stage of agriculture fields from emergence through senescence, vegetation vigor, and other phenomenology.

You can select regions of the Tasseled Cap result in the scatter plot to identify areas of interest.

5. On the **Map** tab, in the **Navigate** group, click **Bookmarks** and select **Heidelberg-Schwetzingen Corridor**.

6. On the **Scatter Plot** chart toolbar, click the **Select Interaction Mode** button and then click **Select by polygon**.

7. On the chart, use your cursor to select the pixels in the upper region of the scatter plot similar to the graphic.

The fields corresponding to the pixel values you selected in the chart are highlighted in the map. The area in the Tasseled Cap distribution represents senescence and the final stage of a crop's life cycle before harvest.

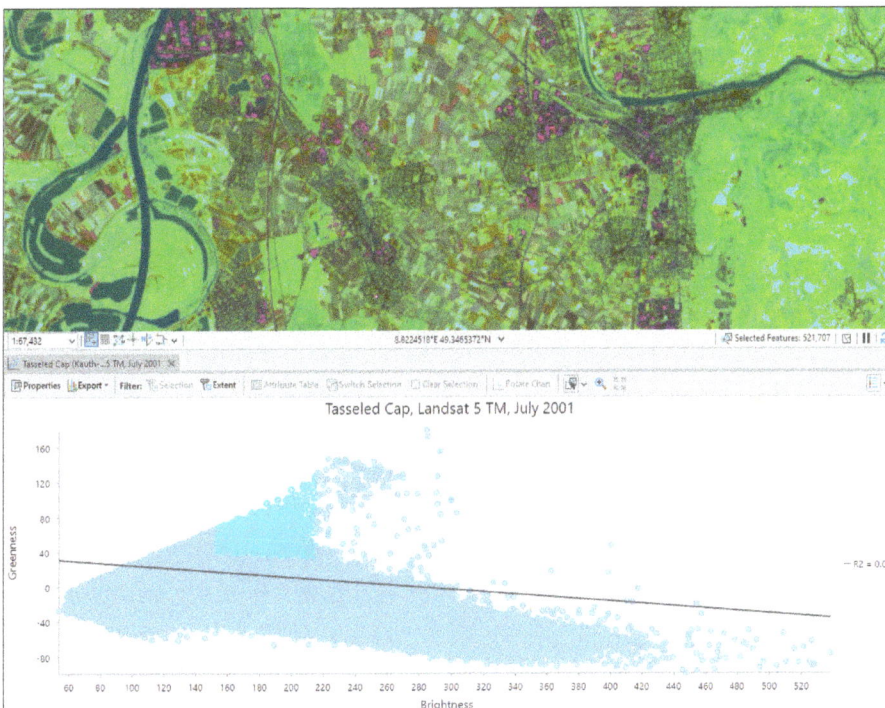

> **Tip:** You can add the **Imagery Hybrid** basemap to your map and in the **Contents** pane move the **Hybrid Reference Layer** to the top of the drawing order to see labels similar to the graphic in this step.

8. On the chart toolbar, export your chart to a format of your choice.

9. Save your project.

Take the next step

- Collect spectral profiles of different areas and examine their properties.
- Create scatter plots of different imagery bands and select features.
- Explore spectral characteristics of various human-made and natural features and see whether you can identify where they would be plotted in a scatter plot.

Summary

In this chapter, you created spectral profiles of areas of interest. You also created scatter plots to examine image bands plotted against each other to understand their spectral characteristics and relationships. Finally, you used the results of a Tasseled Cap transformation in a scatter plot to identify various agricultural fields.

Information at your fingertips

Scatter plots

When deciding which imagery bands to use in scatter plots, it's best to avoid bands that are correlated because they cannot be used to reliably separate multiple features and materials from imagery. Scatter plots can help identify the best uncorrelated bands to use for identifying and extracting features from multispectral imagery. This was why the Blue band and the Near-Infrared band were used in the tutorial.

Things to keep in mind when evaluating the trend line and R2 results from the linear regression:

- When small x-values correspond to small y-values, and large x-values correspond to large y-values (the line sloping up), this indicates a positive correlation. When small x-values correspond to large y-values, and large x-values correspond to small y-values (the line sloping down), this indicates a negative correlation.
- The statistical strength of the correlation is indicated by the R2 value.

The Tasseled Cap (Kauth-Thomas) transformation

Historically, the Tasseled Cap transformation provided a rationale for the patterns found in Landsat Multispectral Scanner (MSS) data of agricultural fields as a function of the life cycle of the crop. Essentially, as crops grow from seed to maturity, there is a net increase in near-infrared and decrease in red reflectance based on soil color. As new sensors became available, the calculations were refined for each sensor expanding from supporting Landsat MSS to include other popular satellite systems. These new sensors include Landsat TM, Landsat ETM+, Landsat 8, IKONOS, QuickBird, WorldView-2, and RapidEye multispectral sensors.

Using the Tasseled Cap function offers several advantages:

- It provides an analytic way to detect and compare changes in vegetation, soil, and human-made features over short- and long-term time periods.
- It provides an analytic way to directly compare land cover features using satellite imagery from different sensors, including Landsat, IKONOS, QuickBird, WorldView-2, and RapidEye.
- It reduces the amount of data from several multispectral bands to three primary components: brightness, greenness, and wetness (what was formally referred to as "yellow stuff" when using Landsat MSS).
- It reduces atmospheric influences and noise components in imagery, enabling more accurate analysis.

References

Crist, E. P., and R. C. Cicone. 1984. "A Physically Based Transformation of Thematic Mapper Data—the TM Tasseled Cap." *IEEE Transactions on Geosciences and Remote Sensing* GE-22: 256–63.

Huang, C. et al. 2002. "Derivation of a Tasseled Cap Transformation Based on Landsat 7 At-Satellite Reflectance." *International Journal of Remote Sensing* 23, no. 1: 741–48.

Kauth, R. J., and G. S. Thomas. 1976. "The Tasseled Cap—a Graphic Description of the Spectral-Temporal Development of Agricultural Crops as Seen by LANDSAT." *LARS Symposia*, paper 159.

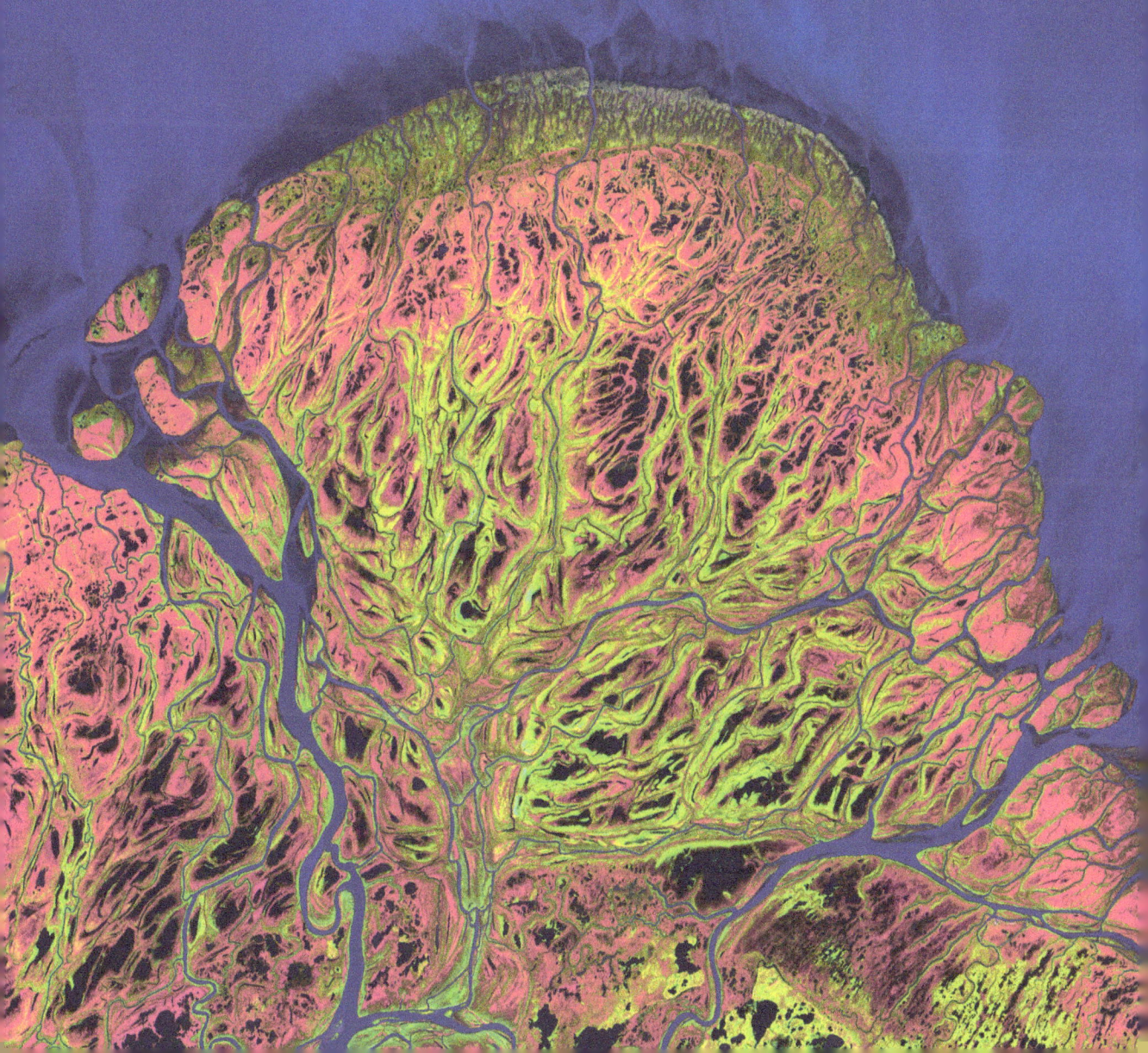

PART 3
Analysis

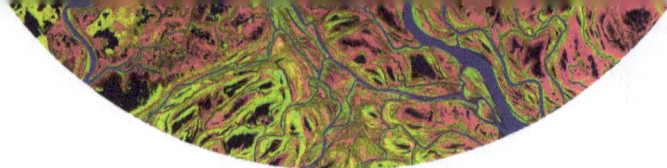

CHAPTER 10
Working with geospatial video

Jeff Liedtke and Tracy Toutant

Objectives

- Work with geospatial video data in the video player.
- Explore the video player controls.
- Use various tools on the video player, including measuring features, creating bookmarks, annotation, video frame capture, video segment capture, and timeline indicators.
- Create a PowerPoint presentation from the video player.

Introduction

Video is widely used across many industries and serves as an asset for applications such as remote inspections and situation awareness. As one of the most current and flexible sources of imagery, video is especially effective for monitoring scenarios that require timely, actionable visual information. With the decreasing cost of video capture, the volume and diversity of video sources continue to expand. ArcGIS is actively evolving to support a broader range of video types within GIS workflows, enabling users to use both the insights derived from video data and the spatial context it provides as part of a comprehensive enterprise GIS.

Tutorial 10-1: Explore geospatial video functionality

Geospatial video—also known as full motion video (FMV) or motion imagery—refers to metadata captured with, or appended to, video content to precisely locate it on a map. With adequate metadata, videos can be synchronized with GIS data, individual frames can be extracted and georeferenced, and data can be collected from the video and overlaid on the video itself and the map bidirectionally. Enabling geospatial capabilities transforms video into a fully integrated component of an enterprise GIS, making it a valuable part of the ArcGIS imagery system.

Download the tutorial data and set up the project

1. Go to links.esri.com/Imagery20Data and download the data for chapter 10.

2. Unzip the folder to **C:\Top20Imagery**.

   > **Note:** In the second chapter, you created a folder named **Top20Imagery** on your drive C. If you haven't done that, create that folder now. Now and in subsequent chapters, you will download and unzip the data for each chapter to this folder.

3. Inside the **Top20Imagery_10** folder, double-click **Top20Imagery_10.aprx** to open the ArcGIS Pro project for this chapter.

Configure ArcGIS Pro and the video player

1. On the ribbon, click the **Project** tab and click **Options**.

2. On the left of the **Options** window, under the **Application** section, click **Full Motion Video**.

3. Expand the **General** section. Check the box for **Enable frame dropping to maintain real-time video playback**.

4. Collapse the **General** section and expand the **Imagery Exports** section. For **Default image format**, click **TIFF**. Check the box for **Add single-frame exports to the map**.

5. Click **OK** to save your settings.

Add a video and use the video player controls

1. In the **Catalog** pane, expand **Folder > Top20Imagery_10**. Right-click the **RedRocks.ts** video and click **Add To Current Map**.

   The **RedRocks** video is listed in the **Contents** pane under the heading **Standalone Videos**, and the video player appears with the video loaded.

   > **Tip:** Click the video player's tab and dock it in a location that also allows you to view the map simultaneously. If you close the video player, you can redisplay it by right-clicking the **RedRocks** video listed under **Standalone Videos** in the **Contents** pane.

2. On the top left of the video player, click the color picker and choose a noticeable color, such as yellow.

3. Click the **Play** button on the video player.

4. As the video plays, right-click on the video in the video player. Click **Zoom To Video** to display the ground track of the drone platform, heading, and the video frame on the map.

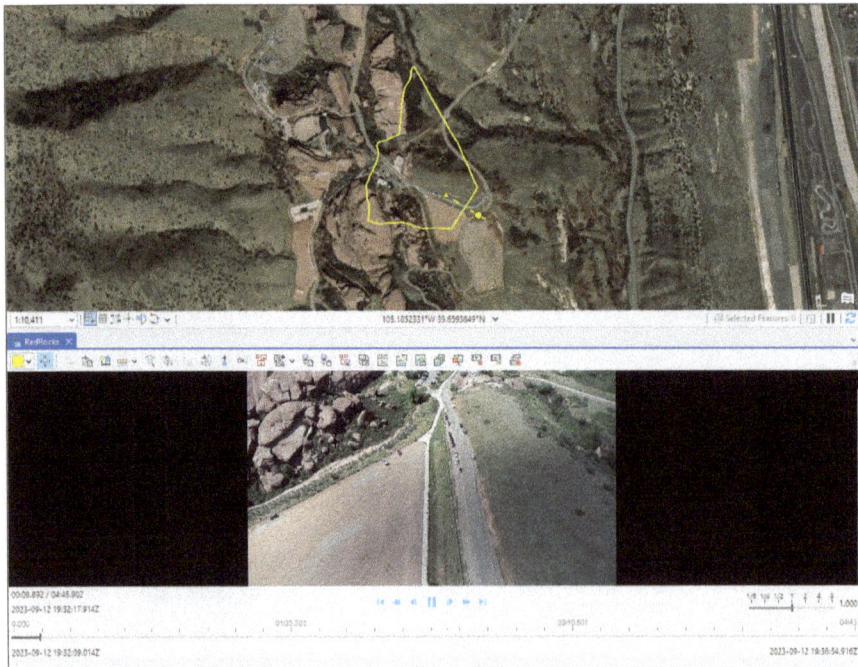

Video footage by Andrew Carey and Alex Posner.[1]

Note: The video footprint looks irregular because it uses a digital elevation model (DEM) to determine the footprint.

5. Pause the video. On the player toolbar, click the **Display Metadata** button.

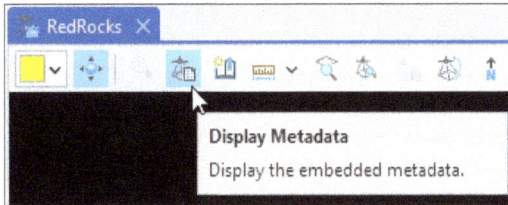

The video display resizes, and the metadata panel appears in the player.

6. Under **Key**, expand **Stream Index 1 > UAS Datalink Local Set** to display the metadata fields and associated values. Observe the types of metadata embedded in the video.

Note: The metadata embedded in the video file enables the display of the video frame, platform trail, position, and heading. This metadata also enables other functionality, such as telestration, measurement of features, bookmarking, and more.

7. Click **Play** and see the metadata being updated in real time as the video plays.

8. Click the **Display Metadata** button again to close the metadata display.

The video display resizes within the player.

9. On the player toolbar, click the Automatic Following button.

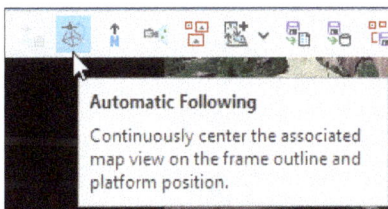

The map pans to keep the sensor and video frame footprint visible in the display. Watch the video and associated positional information update in the map display and video player.

10. Pause the video and note the various player controls and information displayed at the bottom of the video player, such as the beginning and end time of the video, the time stamp of the current video frame, and the playback speed slider. Fast-forward using the buttons and speed slider and click the play button to go back to 1x speed. Try rewind, pause, and step.

11. Click **Play** and use your mouse wheel to zoom in the video player window.

    You can roam the zoomed video in the video player as it plays. The zoom is centered on the cursor in the video player.

12. Click the **Jump To End** button.

Measure features visible in a video frame

You will measure features related to the solar panel installation at the end of the video. To measure features in the video player, do the following steps.

1. On the video player toolbar, click **Measure** and then click **Measure Distance**.

    The pop-up for measuring appears.

2. Click the list in the measure results field and click **Feet**.

    The measurement cursor is displayed in the player.

3. On the video, measure a few rows of solar panels. Click once to start measurement and double-click to end the measurement.

Video footage by Andrew Carey and Alex Posner.

The length is displayed in the video and measurement dialog box.

4. On the video player toolbar, click **Measure** and then click **Measure Area** and measure an area of the solar panels.

5. Close the measurement window.

Collect video bookmarks

Video bookmarks can be collected in the various modes of playing a video, such as play, pause, fast-forward, and rewind. Bookmarks are associated with a time stamp from the active video and can be spatial (2D or 3D) or temporal in the map.

1. Grab the video player progress handle and slide it to the time stamp **02:19**.

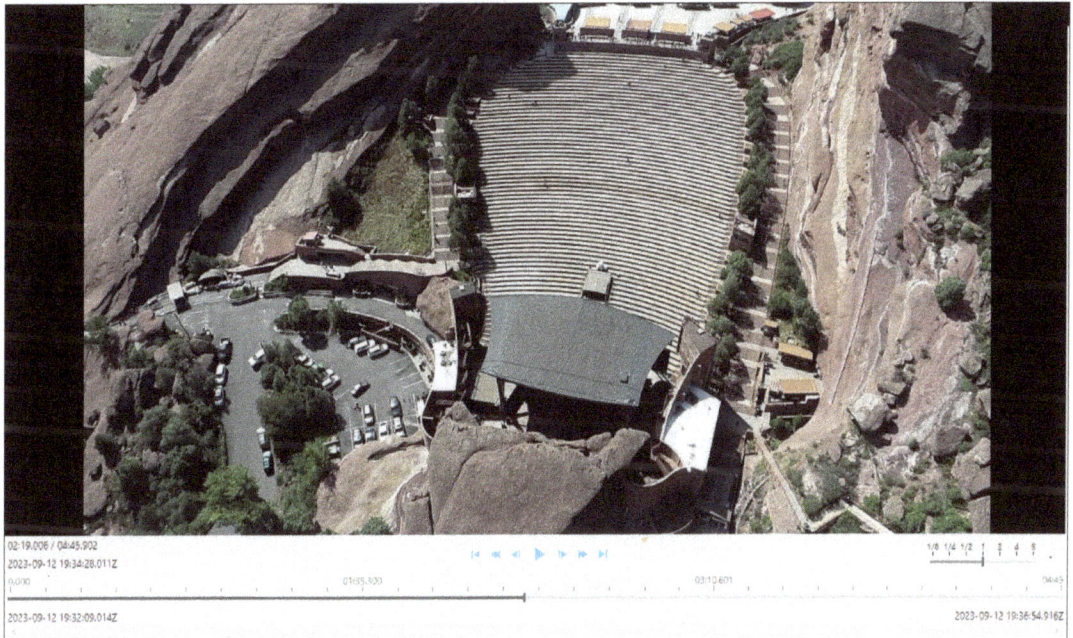

Video footage by Andrew Carey and Alex Posner.

Red Rocks Amphitheatre is displayed in the player.

2. On the player toolbar, click the **Create Bookmark** button. In the **Create Bookmark** dialog box, enter the following settings:

   • **Name**: Amphitheatre
   • **Description**: Red Rocks Amphitheatre

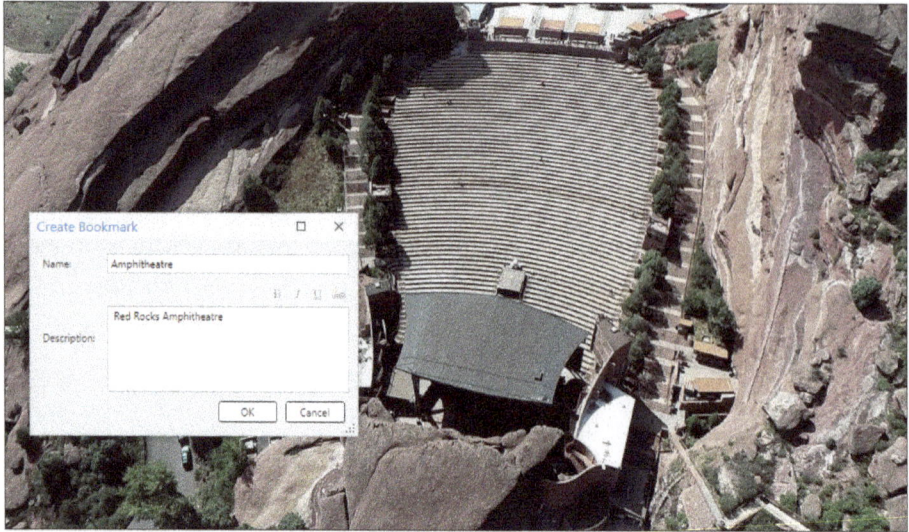

Video footage by Andrew Carey and Alex Posner.

3. Click OK.

   The bookmark and thumbnail overview are created and added to a list.

4. Grab the video player progress handle and slide it to the time stamp **00:27**.

   The Redocks Trading Post is displayed in the player.

5. Click the **Create Bookmark** button. In the dialog box, apply the following settings:

   - **Name**: Trading Post
   - **Description**: Red Rocks Trading Post

6. Click **OK**.

   Bookmarks are managed and selectable in the **Bookmarks** pane. Bookmarks are saved to the project.

7. Click anywhere on the map display. On the **Map** tab, in the **Navigate** group, click **Bookmarks**.

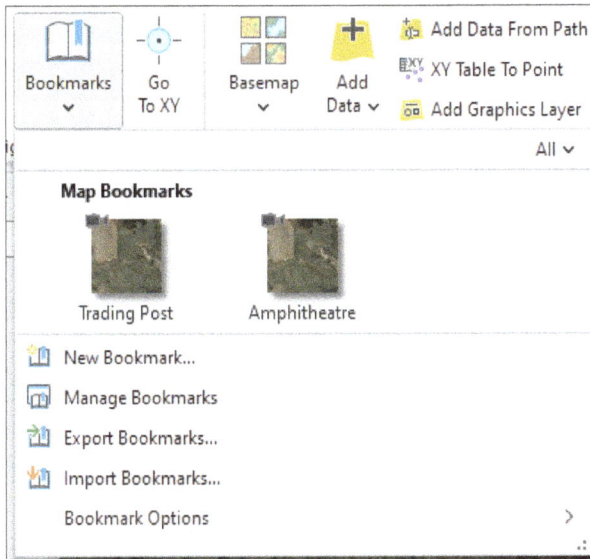

The bookmarks you collected are displayed with the name and thumbnail of each bookmark.

8. Click a bookmark.

The video player jumps to the amphitheater, and the map display is also updated.

9. Click anywhere on the map and the **Bookmark** pane closes.

Create and display annotations

Geospatial video can be annotated using tools that allow you to digitize features on the video and see them displayed on the map within the video footprint. You can also digitize features on the map and see them displayed on the video as it plays.

1. With the video paused at the **Amphitheatre** bookmark, click the **Automatic Dynamic Range Adjustment** button on the video player toolbar.

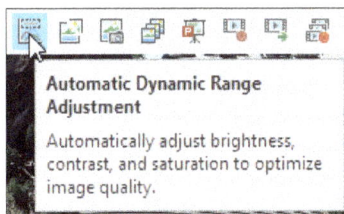

2. On the video player toolbar, click **Create and Display Annotations** list and then click **Annotate Polyline**.

3. On the video player, in the amphitheater, locate the darker colored aisle in the middle of the seating area.

    Tip: Zoom in to the seating area to better see the aisle.

4. Digitize the aisle and double-click to end the polyline.

Video footage by Andrew Carey and Alex Posner.

5. Click **Create and Display Annotations** and then click **Annotate Polygon**.

6. Draw a polygon around the entire amphitheater. Double-click to end the polygon.

Video footage by Andrew Carey and Alex Posner.

The polyline and polygon are displayed in the map.

7. Play the video and see the digitized polygon continue to be displayed on the map and in the video player.

> **Note:** Features created in the video player can be edited, managed, and saved similar to standard features in ArcGIS Pro. Digitized features may be offset because of extreme oblique angles or metadata discrepancies.

8. Pause the video. Click the map to activate it. On the **Map** tab, in the **Selection** group, click **Select** and then click the **Rectangle** selection tool. Select the annotated graphics in the map and then press **Delete** on your keyboard to delete the graphics.

Export the video frame

Next, you will capture and export a video frame.

1. On the **Map** tab, open the **Bookmarks** list and select the **Trading Post** bookmark.

2. On the video player toolbar, click the **Export Frame** button. Click **Play**.

   The video image is displayed on the map.

3. Let the video play for several seconds and then click **Export Frame** again.

4. Let the video play until about **1:25** and then click the **Export Frame** again.

   The exported video images will be displayed on the map.

5. Pause the video player.

Record and export a video segment

Now you will record a segment of an existing video.

1. Move the video progress slider to a location in the video where you want to start recording.

2. On the video player toolbar, click **Record Video**.

3. Let the video play for several seconds, and then pause the video and click the **Record Video** button again to stop the recording.

The **Save Recording** dialog box appears.

4. In the **Catalog** pane, navigate to your project folder and specify a name for your new video file segment. Click **Save**.

5. In the **Catalog** pane, navigate to the **Top20Imagery_CH10** folder. Right-click and refresh the listed contents. Add the new video to the map.

   A video player with your new video appears and is listed under **Standalone Videos** in the **Contents** pane. A default color is assigned to the video player to differentiate it from the **RedRocks** video. Change the color if desired.

6. Play the new video and click the **Automatic Following** button on the player toolbar. After exploring the video, minimize the video player.

Examine and use timeline indicators

A series of colored dots lies below the video timeline in the video player.

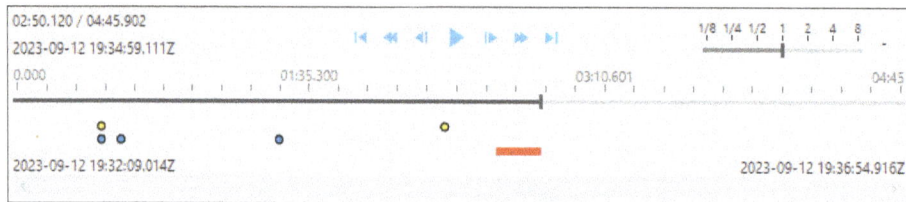

1. Right-click a dot and click **Display Bookmark Indicators** to turn them off.

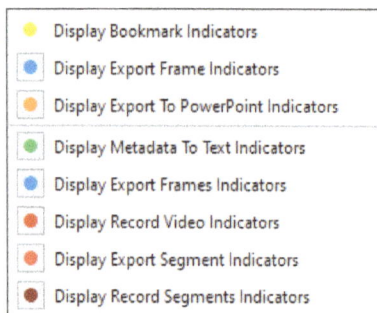

The various indicators listed help you keep track of what processes you have already performed on the video. In this example, there were two yellow book-mark indicators that you turned off, there are three blue indicators where you exported frames and added them to the map, and the red line shows the duration and where in the timeline you recorded a video.

2. After exploring the timeline indicators, minimize the video player.

Export to PowerPoint

Export video content to a PowerPoint presentation by doing the following steps.

1. Go to the **Amphitheatre** bookmark.

2. Activate the video player. On the video player toolbar, click **Export to PowerPoint**.

   The **Export to PowerPoint** dialog box appears.

3. Click the **Browse** button. The **Target PowerPoint Presentations** dialog box appears. Navigate to the **Top20Imagery_10** folder. Type a name for your PowerPoint file and click **Save**.

4. Click **Slide Template** and then click **EsriSampleTemplate1.pptx**. Click **OK**.

   The PowerPoint presentation is created, including the video frame from the player and the map display.

5. In PowerPoint, click the **View** tab and click the **Notes Page** presentation view.

   The video frame and the map view are displayed along with the associated metadata in the Notes pane.

6. In ArcGIS Pro, play the video and click the **Export to PowerPoint** button again. The video frame is captured and saved to the PowerPoint presentation.

   Congratulations! You have a Red Rocks Amphitheatre souvenir, courtesy of geospatial video.

Summary

In this chapter, you used video player controls and tools to display, manipulate, and analyze geospatial video data.

Information at your fingertips

Geospatial video basics

Geospatial video combines traditional video files or streams with metadata about the collection of the video for georeferencing. There are many ways to collect video, including satellites, commercial drones, full motion video, handheld cameras, mobile and stationary mounted cameras, and more. All these methods contain a variety of ways to capture metadata about the video. Some contain little to no reference information, some have an external file with information about location, and some systems collect detailed information on location, altitude, and camera orientation. ArcGIS uses what metadata is available to locate the video on the map. The more metadata provided, the more functionality becomes available, but even videos with no metadata can be played in the video player.

Multiplexing

The process of combining video and metadata is called multiplexing. Multiplexing encodes the available metadata information into the video file so that each frame in the video has geospatial information associated with it.

When video is geospatially enabled, ArcGIS allows you to see the available metadata on the map while the video is playing. This may be a static point location of a stationary camera, a moving point of a dashcam, or a set of features that represent the location of the collection platform with its full footprint of the video. If you have enough metadata available, you can also view your GIS data overlaid on the video and create features on the video that also appear in the map view.

See links.esri.com/RedRocks for an ArcGIS StoryMaps[SM] story about how the Red Rocks video was collected and processed.

See links.esri.com/MultiplexerTips for information on how to combine a video file with its associated metadata file.

See links.esri.com/Geospatial_Video_Player for a comprehensive description of the ArcGIS Pro geospatial video player. This link takes you to the ArcGIS Pro reference help. In the table of contents, you will see several help topics describing geospatial video capabilities, such as searching video archives, tracking objects in the video, and more.

Note

1. Video footage shot at the Red Rocks Amphitheatre courtesy of the City and County of Denver Technology Services.

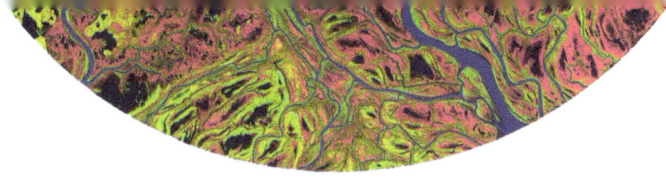

CHAPTER 11
Performing change detection

Hong Xu

Objectives

- Explain change detection methods.
- Detect land cover change using two land cover datasets.
- Explore time series change interactively.
- Create a disturbance map using a Landsat time series.

Introduction

Change detection is essential for understanding and monitoring environment trans-formation over time, providing valuable insights for informed decision-making. It involves comparing two or multiple images collected at different times to identify areas of change between those time points. With the availability of large archives of remote sensing imagery, such as 50-plus years of Landsat imagery, change detection has evolved to include the automatic detection of change points and the extraction of detailed change information. This process includes identifying when and where change occurred, its characteristics (such as increasing or decreasing trends), the magnitude of the change, and the duration of the transformation.

ArcGIS provides tools and functions for two typical workflows:

- **Image to image change detection**: This approach involves comparing pixel value changes, spectral angle differences, or categorical change, such as land cover.
- **Time series change detection**: This approach supports advanced algorithms, including LandTrendr (Landsat-based detection of trends in disturbance and recovery) and continuous change detection and classification (CCDC).

Some change detection workflows require preprocessing, such as classifying images with machine learning or deep learning approaches, or creating a band index using raster functions.

This chapter focuses on two change detection tutorials. First, you will learn how to use the Change Detection Wizard tool and the National Land Cover Database (NLCD) to analyze and quantify land cover change. The second tutorial detects forest disturbance using a Landsat image time series.

Tutorial 11-1: Detect land cover change using two NLCD datasets

The USGS has released NCLD products for the conterminous United States across nine epochs from 2001 and 2021, enabling analysis of the land cover change between any two epochs. In this tutorial, you will use the 2001 and 2021 NLCD datasets to analyze land cover change in Austin, Texas, and its surrounding areas over a 20-year period. You will learn how to work with the Change Detection Wizard tool to compute and map these changes and then use charts to visualize and analyze the results effectively.

Download the tutorial data and set up the project

1. Go to links.esri.com/Imagery20Data and download the data for chapter 11.

2. Unzip the folder to **C:\Top20Imagery**.

   Note: In the second chapter, you created a folder named **Top20Imagery** on your C: drive. If you haven't done that, create that folder now. Now and in subsequent chapters, you will download and unzip the data for each chapter to this folder.

3. Inside the **Top20Imagery_11** folder, double-click **Top20Imagery_11.aprx** to open the ArcGIS Pro project for this chapter.

    Two images, **NCLD_Austin_2001.tif** (*left*) and **NCLD_2021.tif** (*right*), and their land classifications are listed in the **Contents** pane.

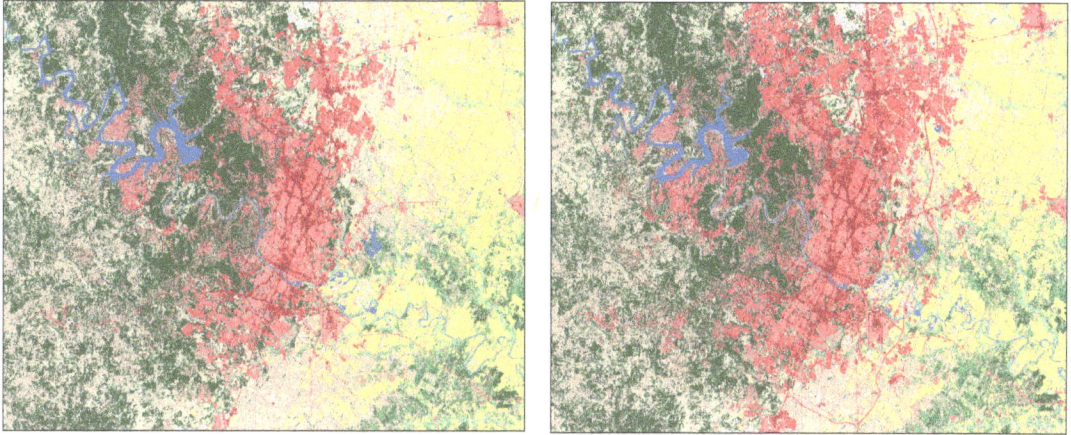

By visually comparing the two datasets, you can observe an increase in developed areas in 2021. The tutorial helps qualify and quantify changes, determining what land cover was converted to development and what land cover had forest loss.

Open the Change Detection Wizard and configure inputs

1. In the **Contents** pane, select the **NCLD_Austin_2001.tif** layer.

2. On the ribbon, click the **Imagery** tab, and in the **Analysis** group, click **Change Detection > Change Detection Wizard**.

3. In the **Change Detection Wizard** pane, for **Configure**, apply the following settings:
    - **Change Detection Method:** Categorical Change
    - **From Raster:** NLCD_Austin_2001.tif
    - **To Raster:** NLCD_Austin_2021.tif

4. Click **Next**.

Filter change categories

The **Class Configuration** section lists all class names for both **From Classes** and **To Classes**. Although there can be many change permutations between these two land cover datasets, some land cover types remain unchanged. This pane enables you to filter these various change types easily. In this section, you will filter the land cover types that are converted to developed land.

1. For **Class Configuration**, confirm that the **Filter Method** is set to **Changed Only**. For **From Classes**, keep all the boxes checked. For **To Classes**, click **Unselect All**. Check the boxes for the following classes:
   - **Developed Open Space**
   - **Developed Low Intensity**
   - **Developed Medium Intensity**
   - **Developed High Intensity**

2. Click **Preview**.

   A layer named **Preview_ComputeChange** is added.

3. Turn off visibility for the two **NLCD** layers below the preview layer to see the change better.

   The preview layer contains all the pixels that have been changed in the four classes.

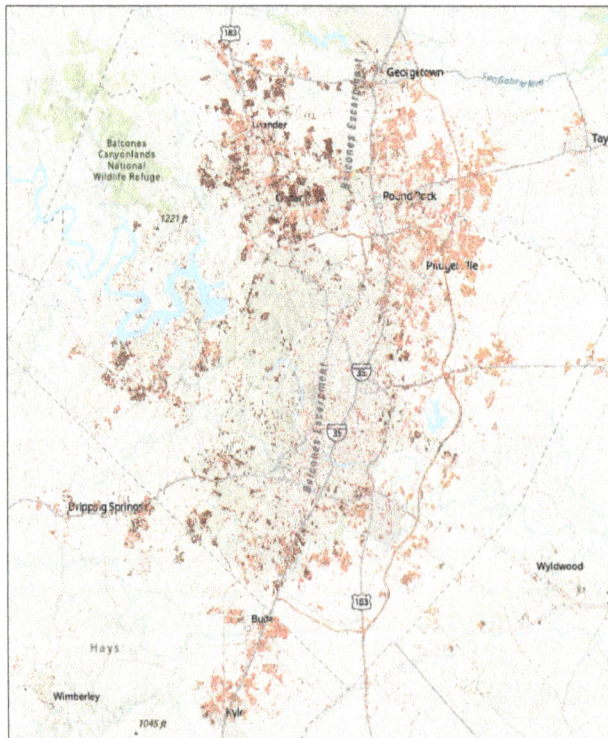

4. In the **Contents** pane, expand the **Preview_ComputeChange** layer.

   The layer's legend depicts all the change categories.

Save and quantify the change

In this section, you will save the detected changes to a TIFF file and analyze the changes using charts.

1. In the **Change Detection Wizard** pane, click **Next**.

2. In the **Output Generation** section, change the name of the **Output Dataset** to land_cover_changes.tif.

3. Click **Run**.

   The **land_cover_changes.tif** raster will be added to the map.

4. Click **Finish** to close the **Change Detection Wizard** pane.

5. In the **Contents** pane, right-click the **land_cover_changes.tif** layer and click **Attribute Table**.

   You will select all the rows in the attribute table except the last two. These two rows are unchanged and will not be included in the chart.

6. In the table, select row 1, press **Shift**, and select row 50.

7. On the top right of the table, click the three lines and click **Export**.

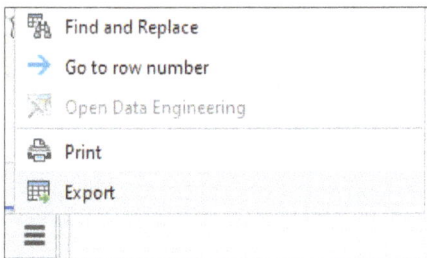

8. In the **Export Table** window, for **Output Table**, change the name to land_cover_changes_export. Click **OK**.

9. Close the table.

10. In the **Contents** pane, under **Standalone Tables**, right-click **land_cover_changes_export**. Hover over **Create Chart** and click **Bar Chart**.

11. In the **Chart Properties** pane, apply the following settings:
    - **Category or Date**: Class_From
    - **Aggregation**: Sum
    - **Numeric field(s)**: Select > Count

The chart is created with all the records that were specified in the table and depicts a summary of land types that changed to developed categories. Many of the changes are from development on grassland and evergreen forests.

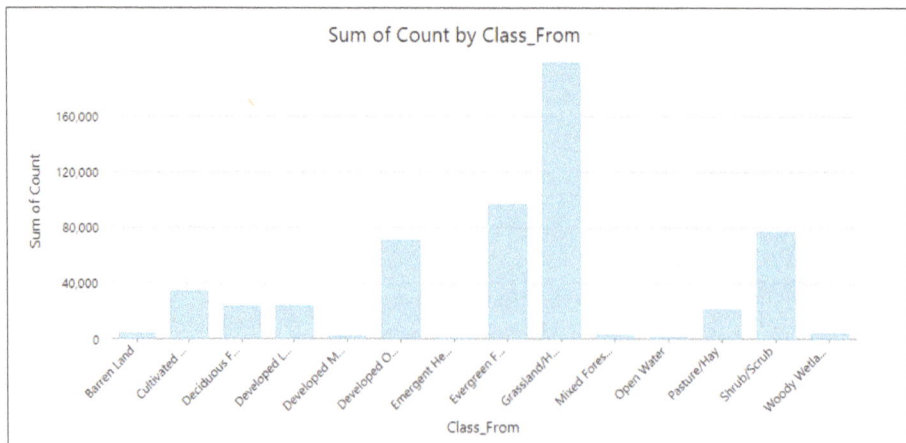

12. Close the chart tab.

Tutorial 11-2: Detect change using a Landsat image time series

With the availability of large archives of satellite imagery, analyzing an image time series provides comprehensive information, such as when and where change has occurred, whether these changes are increasing or decreasing, the magnitude of the change, and the duration of the change events.

Forests are a vital natural resource, offering wood products, wildlife habitat, carbon sequestering, and more. Events such as logging, forest fires, and pest infestations significantly impact forest growth. Monitoring and mapping these disturbances are crucial tasks for effective forest management. In this tutorial, you will focus on a region in the West Cascades of Oregon to learn about image time series change detection using the LandTrendr method. You will begin by getting familiar with the Landsat image cube, which stores the Landsat image time series. Next, you will create a normalized burn ratio (NBR) image cube data model, learn how to interactively explore and analyze forest disturbance within an image cube, and then map the disturbance events.

Get familiar with the input image cube

The input image cube is created from the Landsat images of the summer season from 1980 to 2020[1].

1. On the ribbon, click the **Insert** tab. In the **Project** group, click **New Map** to add a new map.

2. In the **Catalog** pane, expand **Folders > Top20Imagery_11**. Right-click the **West_Cascade.crf** image cube and click **Add To Current Map**.

3. Select the layer in the **Contents** pane and click the **Multidimensional** tab.

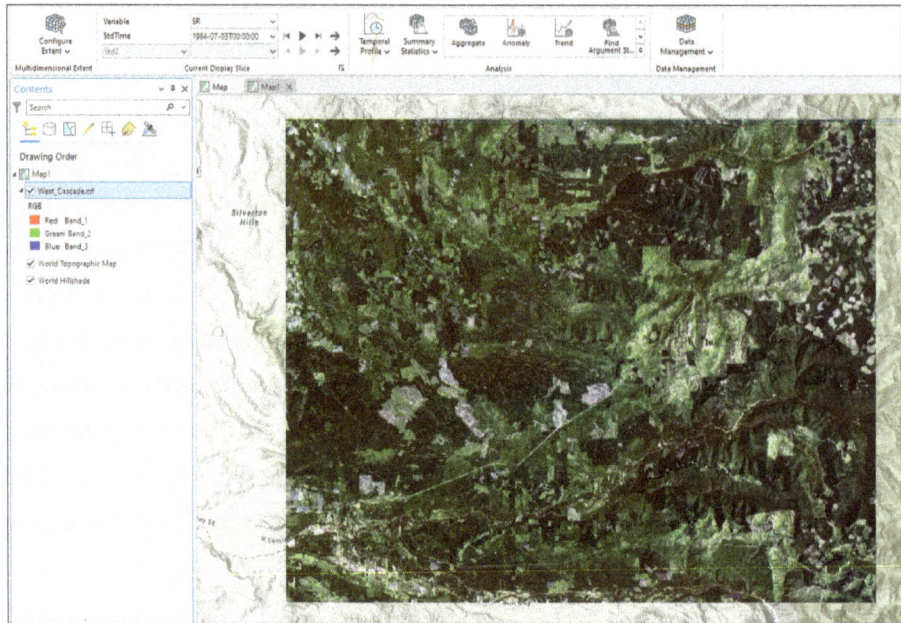

4. In the **Current Display Slice** group, click **StdTime** and review the list of time dimension values from 1984-7-3 to 2020-8-3, representing the dates these image slices were captured, from July 3, 1984, to August 3, 2020.

5. Select the **2011-07-30T00:00:00** time slice to display it on the map.

    The image cube contains images in the summer for each year. Some years do not
    have images because no usable images are available.

Create an NBR image cube

The LandTrendr application operates on a single band of the image cube. The NBR
commonly used for mapping forest burn scars is also effective for distinguishing
healthy forest from forest disturbances. Next, you will create an NBR image cube
from the Landsat image cube.

1. In the **Contents** pane, make sure the **West_Cascade.crf** layer is selected.

2. On the ribbon, click the **Imagery** tab. In the **Tools** group, click **Indices**. Under
    the **Landscape** group, click **NBR**.

3. In the **NBR** dialog box, specify the **Near Infrared Band Index** as **Band_4** and
    the **Shortwave Infrared Band Index** as **Band_6**.

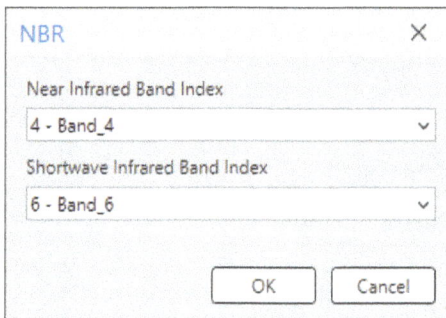

4. Click **OK**.

    A new layer named **NBR_West_Cascade.crf** is added to the map and listed in
    the **Contents** pane. The output layer is an NBR data cube computed on the fly
    using the **Band Arithmetic** function with the NBR required bands.

5. In the **Contents** pane, under the **NBR_West_Cascade.crf** layer, right-click the
    legend, expand the list, and choose the **Pink-Green (Continuous)** color ramp.

    Tip: In the list, click **Show Names** to find the color ramp.

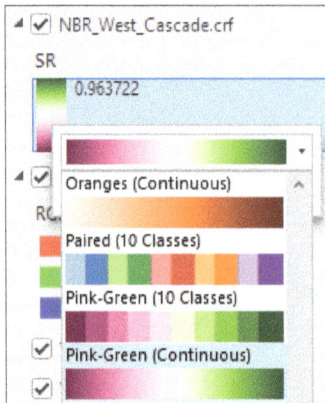

The map is updated to reflect the pink and green color scheme.

From visual examination, pixels in pink are logging disturbances and pixels in green are forest.

6. Click the **Multidimensional** tab. In the **Current Display Slice** group, for **StdTime**, click the **Step Forward** button (right arrow) to observe the changes over time for the whole area of interest.

7. Set the current time slice to **2017-07-14T00:00:00**.

Explore a specific pixel's change over time

You will use the **Pixel Time Series Change Explorer** tool to explore the forest change at one location at a time.

1. In the **Contents** pane, make sure the **NBR_West_Cascade.crf** image cube layer is selected.

2. On the **Multidimensional** tab, in the **Analysis** group, click **Temporal Profile** and then click the **Pixel Time Series Change Explorer** tool.

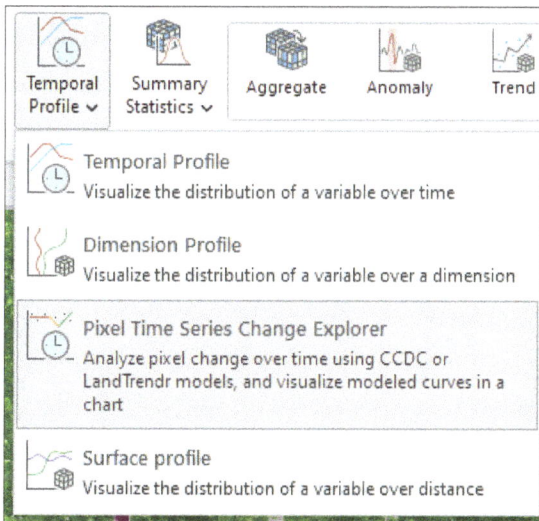

The **Chart Properties** pane appears, and an empty chart appears below the map.

3. In the **Chart Properties** pane, set the **Change detection method** to **LandTrendr**.

4. Under **Define a pixel location**, click the **Point** button. Click on a logged (green) area on the map.

For example, try the location 549,326E 4,970,700N m, as shown on the next page.

> **Tip:** On the **Map** tab, in the **Navigate** group, use the **Go To XY** tool and enter the coordinates using the **Meters** option.

5. Expand the **Model Parameters** section. Change the **Maximum Number of Segments** to 4 to ignore small changes.

6. Click **Fit and create chart**.

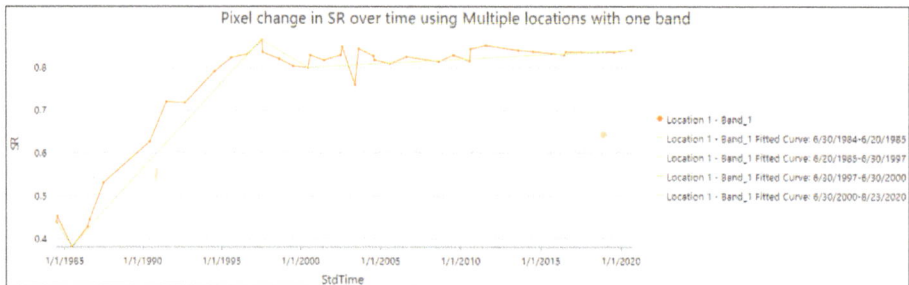

The orange line represents NBR observations over time. Smaller differences may result from differences in image quality or acquisition dates rather than actual changes. A sudden decrease is evident, followed by a gradual recovery.

7. On the chart toolbar, click **Select interaction mode** and then click **Select by rectangle**. Highlight different time points and observe the map to see the corresponding NBR image at that time.

8. Explore other locations of interest, and when you are finished, close the chart window.

Analyze and create a change analysis raster

Next, you will use the **Analyze Change Using LandTrendr** tool to analyze the change for all pixels in the image and create a change analysis raster. This change analysis raster is an image cube that stores the LandTrendr regression model. Each slice is a multiband raster containing model coefficients, such as slope, intercept, and fitted values. If no change is detected, the slices remain the same. This change analysis raster provides rich information for deriving types of change across the study area.

1. In the **Contents** pane, make sure the **NBR_West_Cascade.crf** layer is selected.

2. On the **Multidimensional** tab, in the **Analysis** group, expand the **Analysis** tool gallery and click the **Analyze Changes Using LandTrendr** tool.

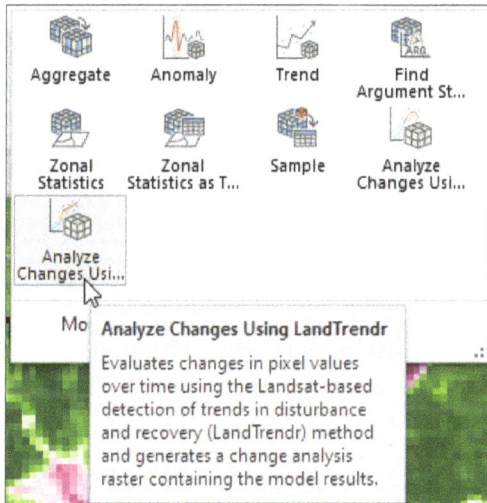

The **Analyze Changes Using LandTrendr** tool appears with its default settings.

3. Change the **Maximum Number of Segments** to 4. Keep all other defaults as is.

4. Click **Run** to generate the change analysis raster. Depending on your machine, this may take time.

    An **NBR_West_Cascade_LandTrendr.crf** layer is added to the map with the first band displayed on the map.

5. On the ribbon, click the **Raster Layer** tab and click **Symbology**.

6. In the **Symbology** pane, set the **Band** to **Band_1_Fitted_Value**.

7. Check the box for **Invert** to invert the color ramp.

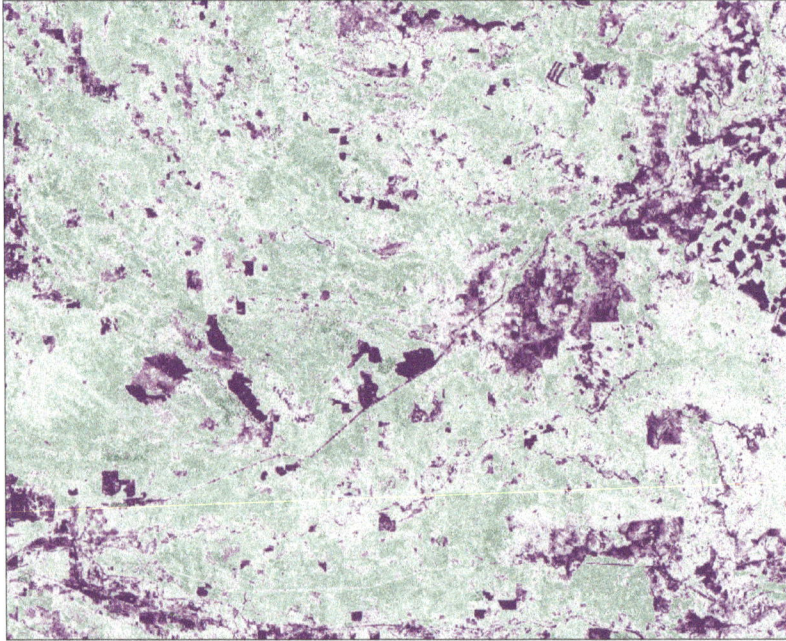

The graphic shows a fitted model of the original NBR image cube, where pink and purple indicate changes over the time span. In addition to the fitted values, it also contains bands for the slope and change magnitude, from which you will extract change properties. You can also use the temporal profile tool to visualize the model.

8. On the **Multidimensional** tab, in the **Analysis** group, click **Temporal Profile**.

9. In the **Chart Properties** pane, click the **Point** button. On the map, click a changed (green) location.

10. In the **Select bands** section, change the **Band** to **Band_1_Fitted_Value**.

The chart now depicts the change occurring at this location. It remains healthy forest until trees are logged and replanted and regrowth occurs.

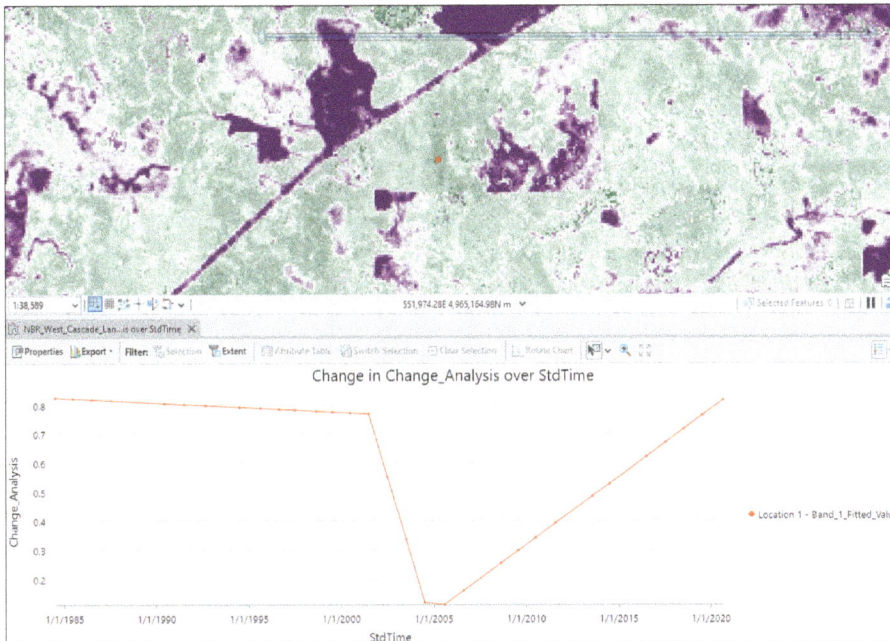

11. Close the chart window.

Extract the change and create a disturbance map

Now that you have a change analysis raster, you can use it to generate informative maps. In this section you will use the **Detect Change From Change Analysis Raster** tool to extract change dates and map the change over time. The logging disturbances can be described as follows:

- NBR values decrease through time.
- Change occurs abruptly, instead of occurring over many years.
- The change magnitude is above a threshold that can filter out small variations.

1. In the **Contents** pane, ensure the **NBR_West_Cascade_LandTrendr.crf** is selected.

2. On the **Analysis** tab, in the **Geoprocessing** group, click **Tools**.

3. In the **Geoprocessing** pane, search for and open the **Detect Change Using Change Analysis Raster** tool.

Because a change includes start and end times and a pixel can have multiple changes, you will extract the end time and thus extract the last change.

4. Apply the following settings for the tool:
   - **Input Change Analysis Raster**: NBR_West_Cascade_LandTrendr.crf
   - **Output Raster**: Disturbance.crf
   - **Segment Date**: End of segment
   - **Change Direction**: Decreasing
   - **Change Type**: Time of Latest Change
   - **Maximum Number of Changes**: 1

5. Under **Filter By Attributes**, check the box for the following attributes and apply the appropriate values:
   - **Filter By Duration – Minimum Duration**: 0, **Maximum Duration**: 3
   - **Filter By Start Value – Minimum Start Value**: 0.7, **Maximum Start Value**: 1
   - **Filter By End Value – Minimum End Value**: –1, **Maximum Start Value**: 0.5

   The values are used to filter out unwanted changes. The starting value range defines healthy forest, and the ending value range defines the disturbance pixels. These are calculated from sample points.

6. Click **Create new layer** to create the disturbance map.

7. In the **Contents** pane, turn off the **NBR_West_Cascade_LandTrendr.crf** and **NBR_West_Cascade.crf** layers.

The pixel values in blue are dates where the disturbance occurred in early years, and red and warm colors mean the events happened in recent years.

Symbolize the raster

To make an appealing map, you will do some post processing to fill holes in the logging plots using a raster function.

1. In the **Raster Function** pane, search for Statistics and open the **Statistics** function.

2. Apply the following settings for smoothing:
    - **Raster**: Disturbance.crf
    - **Statistics Type**: Majority
    - **Number of Rows**: 3
    - **Number of Columns**: 3
    - Check the box for **Only fill NoData pixels**

3. Click **Create new layer**.

    This function fills the NoData pixels within a 3 by 3 neighborhood with the majority values and creates a smoother map. To finish the map, you will define your preferred color ramp and labels.

4. On the **Raster Layer** tab, click **Symbology**.

5. In the **Symbology** pane, choose a color ramp and modify the date labels to produce an attractive map.

The disturbance map is just one of the maps that can be extracted from the change analysis raster. You can also extract a map showing the recovery processes or classify the forest types using the change analysis raster[2].

Summary

In this chapter, you learned two typical change detection workflows: image-to-image change detection and image time series change detection. You used two NLCD datasets to analyze land cover changes between 2001 and 2021 in Austin, Texas, and applied the Landsat image time series to do the interactive change detection workflow. Additionally, you detected change and created a disturbance map from a Landsat image time series. Using the Change Detection Wizard, Pixel Time Series Change Explorer, and the change detection geoprocessing tools, you applied these techniques to your areas of interest.

Information at your fingertips

The two datasets used for this exercise have been prepared. If you are interested in other areas, the NLCD product can be accessed from multiple places, including the following:

- **USGS Earth Explorer** at links.esri.com/EarthExplorer
- **Microsoft Planetary Computer** at links.esri.com/MPC
- **ArcGIS Living Atlas** at links.esri.com/LivingAtlas

Notes

1. Create Image Cubes from STAC-Enabled Datasets using ArcGIS at links.esri.com/CreateImageCube.
2. Monitor forest change over time at links.esri.com/MonitorForestChange.

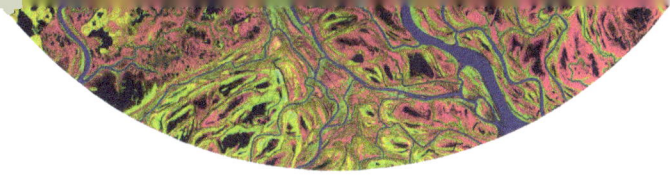

CHAPTER 12
Performing multispectral classification

Objectives

- Explore the multispectral classification process.
- Address the essential skills needed for each step of the process, including image segmentation, classification schema, collection of training samples, training the classifier, performing classification, merging classes, accuracy assessment, remapping classes, and updating classification.
- Visualize and analyze imagery, raster data, and supporting information for classification.
- Explore techniques and best practices to produce credible class maps.

Introduction

Image classification is the art and science of quantifying the identification of features or objects in imagery. It uses the spectral, shape, and spatial characteristics of objects in imagery to categorize them into representative classes. The resulting class map is used as an input to GIS modeling, land/asset inventory and management, and multiple levels of decision support.

You will use the object-oriented classification approach, where groups of similar pixels are combined into segments. Segments exhibiting certain shapes and spectral and spatial characteristics can be further grouped into objects. The objects can then be grouped into classes that represent real-world features on the ground. This is referred to as object-based image analysis (OBIA).

In this chapter, you'll learn to classify a multispectral image and assess the feature accuracy of the classified map. You will use the classification wizard, which guides the workflow, starting with configuring the project, collecting class training samples according to a defined schema, training and classifying the imagery, merging and assigning classes, and accuracy assessment. By the time you complete this chapter, you'll be prepared to implement these techniques in real-world applications.

Tutorial 12-1: Perform multispectral classification and segmentation

In this tutorial, you will use multispectral aerial imagery to create an impervious surface map, used to monitor pollution from stormwater runoff, determine stormwater utility fees, or plan flood control and emergency management strategies.

You will create an impervious surface map (ISM), where human-made features that prevent water—primarily precipitation—from soaking into the ground will be identified and classified.

Download the tutorial data and set up the project

1. Go to links.esri.com/Imagery20Data and download the data for chapter 12.

2. Unzip the folder to **C:\Top20Imagery**.

> **Note:** In the second chapter, you created a folder named **Top20Imagery** on your C: drive. If you haven't done that, create that folder now. Now and in subsequent chapters, you will download and unzip the data for each chapter to this folder.

3. Inside the **Top20Imagery_12** folder, double-click **Top20Imagery_12.aprx** to open the ArcGIS Pro project for this chapter.

4. In the **Catalog** pane, expand **Folders > Top20Imagery_12**. Right-click **Charlotte_NAIP2023.crf** and click **Add To Current Map**.

    The image data is listed in the **Contents** pane and loaded into the map.

5. Explore the image data by roaming, zooming, and switching the band combination.

6. In the **Contents** pane, right-click **Charlotte_NAIP2023.crf** and click **Properties**.

7. In the **Layer Properties** window, click **Source**. Click the expander arrows for the different properties.

   Note that the imagery comprises four bands, ordered as Red (band 1), Green (band 2), Blue (band 3), and Infrared (band 4).

8. Close the **Layer Properties** pane.

Generate a segmented image

The first step in the image classification workflow using the object-oriented approach is to generate a segmented image. Image segmentation is a computationally intensive process using a geoprocessing tool, which produces an output file saved to disk. To determine the optimum parameter settings, you will first use the segmentation raster function, called **Segment Mean Shift**, to experiment with various parameter and option settings best suited for the NAIP imagery. Then you will use those settings to run the **Segment Mean Shift** geoprocessing tool to produce a saved segmented image.

You will set the image bands to Infrared, Green, and Blue.

1. In the **Contents** pane, under the **Charlotte_NAIP2023.crf** layer, right-click the **Red** display band and click **Infrared**.

2. Right-click the **Green** display band and click **Red**.

   The **Blue** band is already loaded into the **Blue** display.

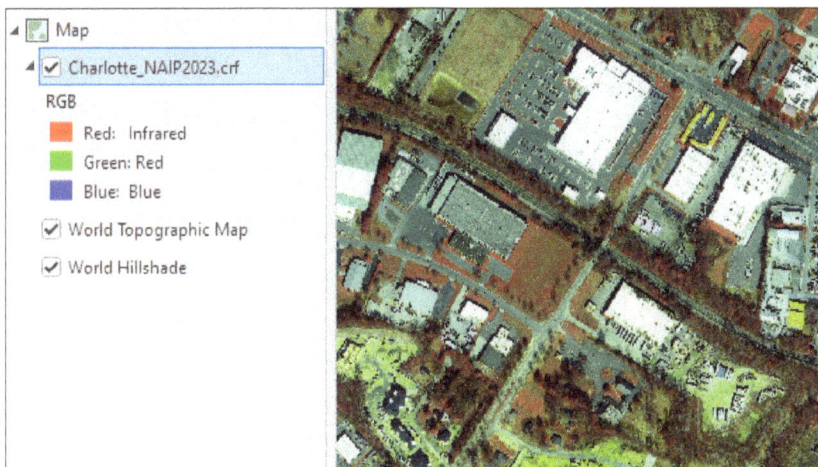

3. On the **Imagery** tab, in the **Analysis** group, click **Raster Functions**.

    The **Raster Functions** pane appears and is docked.

4. At the top of the **Raster Functions** pane, click the search bar and type Extract. Click the **Extract Bands** raster function.

5. In the **Extract Band Properties** function, apply the following settings:
    - **Raster**: Charlotte_NAIP2023.crf
    - **Method**: Band IDs
    - **Combination**: 4 1 3

    Tip: Segmentation works on three-band imagery only. Choose the three image bands that best distinguish the features of interest. When typing the band combination, ensure that you keep a space between each number.

6. Click **Create new layer**.

    The layer **Extract Bands_Charlotte_NAIP2023.crf** is listed in the **Contents** pane and added to the map.

Use the Segment Mean Shift raster function

1. In the **Raster Functions** pane, search for segment and click the **Segment Mean Shift** raster function.

2. In the **Segment Mean Shift** function, for **Raster**, select the **Extract Bands_ Charlotte_NAIP2023.crf** layer you created in the previous step. Click **Create new layer**.

The segmented image, named **Segment Mean Shift_Extract Bands_Charlotte_ NAIP2023.crf**, is created and listed in the **Contents** pane and added to the map.

3. Create two additional **Segment Mean Shift** raster function layers using different parameters, as follows:

   - **Raster**: Extract Bands_Charlotte_NAIP2023.crf
   - **Spectral Detail**: 10
   - **Spatial Detail**: 8
   - **Raster**: Extract Bands_Charlotte_NAIP2023.crf
   - **Spectral Detail**: 18
   - **Spatial Detail**: 8

4. Explore the various segmented images by turning them off and on in the **Contents** pane.

5. On the **Raster Layer** tab, in the **Compare** group, use the **Swipe** tool as another way to view the images.

   The level of detail varies, depending on the different spectral and spatial detail settings.

Create the saved segmented image

Creating the segmented image is a computationally intensive process and may take about 35 minutes to complete.

> **Tip:** You can skip this section and use the segmented image provided in the project folder, **SegmentedCharlotteNAIP_413.crf**.

1. On the **Analysis** tab, in the **Geoprocessing** group, click **Tools**.

2. In the **Geoprocessing** pane, search for and open the **Segment Mean Shift** geoprocessing tool. Apply the following settings:
   - **Input Raster**: Extract Bands_Charlotte_NAIP2023.crf
   - **Output Raster Dataset**: SegmentedCharlotteNAIP.tif
   - **Spectral Detail**: 17
   - **Spatial Detail**: 7
   - **Minimum Segment Size In Pixels**: 20
   - **Maximum Segment Size In Pixels**: –1

3. Click the **Environments** tab. Set the **Parallel Processing Factor** to 80%.

4. Click **Run**.

    The segmented image will take time to create and will be saved in the output folder specified.

5. On the map, explore the segmented image **SegmentedCharlotteNAIP.tif**.

Tutorial 12-2: Use the Training Samples Manager

In this tutorial, you will collect training samples that define the impervious and pervious features in the imagery. These training samples will be used to train the classifier.

1. In the **Contents** pane, turn off all layers except **SegmentedCharlotteNAIP.tif**. Select the layer.

2. Click the **Imagery** tab. In the **Image Classification** group, click **Classification Tools** and then click **Training Samples Manager**.

    The **Image Classification** pane appears. You will see the schema management section at the top, where the default schema (**NLCD2011**) is automatically loaded.

    > **Note:** Establishing a classification schema and collecting training samples is important for the accuracy and success of your classification project. Collecting representative training samples can be time consuming. A classification schema and training samples file is provided in the tutorial data and can be loaded in the **Training Samples Manager**.
    >
    > If you choose to use the existing classification schema and training samples, it is highly recommended that you collect several additional training samples for each class according to the steps detailed here.

3. In the **Image Classification** pane, click the **Create New Schema** button.

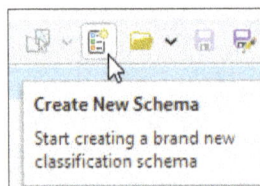

A new entry named **New Schema** is added to the pane.

4. Right-click the **New Schema** heading and click **Add New Class**.

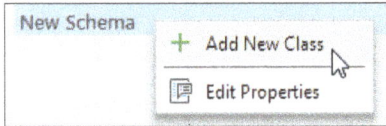

5. In the **Add New Class** pane, apply the following settings:
   - **Name**: Impervious
   - **Value**: 10
   - **Color**: Gray 40%

6. Click **OK**.

   The window closes and the parent class **Impervious** is added to the schema.

7. For **New Schema**, add another parent class. Name the class Pervious, set the **Value** to 60, and change the **Color** to **Leaf Green**. Click **OK**.

   The window closes and the parent class **Pervious** is added to the schema.

8. Right-click the **Impervious** parent class and click **Add New Class**. Add the following subclasses:
   - **Name**: Asphalt, **Value**: 15, **Color**: Gray 50%
   - **Name**: Concrete, **Value**: 20, **Color**: Gray 10%
   - **Name**: White Roofs, **Value**: 25, **Color**: Arctic White
   - **Name**: Gray Roofs, **Value**: 35, **Color**: Gray 40%

9. Right-click the **Pervious** parent class and click **Add New Class**. Add the following subclasses:
   - **Name**: Grass Turf, **Value**: 65, **Color**: Quetzal Green
   - **Name**: Bare Trees, **Value**: 70, **Color**: Tarragon Green
   - **Name**: Leafy Trees, **Value**: 75, **Color**: Spruce Green
   - **Name**: Bare, **Value**: 80, **Color**: Raw Umber

10. Click the **New Schema** heading. Click **Edit Properties** and update the name to ImperviousSchema. Click **Save**.

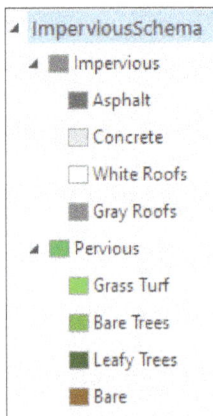

11. In the **Image Classification** pane, click the **Save As** button.

12. For **Output Location**, click the **Browse** button and save the schema to **C:\ Top20Imagery\Top20Imagery_12\Training**. Click **OK** to close the dialog box.

13. Click **Save**.

Collect training samples

Collect several training samples for each class. The training sample polygons will comprise segments derived from the image.

1. Zoom in to an area of the image so that several asphalt features are visible.

2. In the **Image Classification** pane, click the **Asphalt** subclass. Click the **Segment Picker** tool and click **SegmentedCharlotteNAIP.tif**.

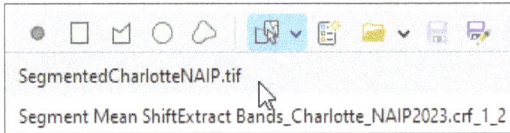

The cursor on the map has the **Segment Picker** icon displayed with the cursor.

3. In the image, click on an asphalt feature to highlight the segment. Click again to capture the asphalt training sample.

4. Click on additional asphalt features in different parts of the image to collect about 20 representative training samples for the **Asphalt** subclass. Collect lighter and darker asphalt features to collect representative asphalt features. Make sure the segment represents the class and does not include other nonasphalt features.

> **Tip:** To roam to other areas in the image, on the **Map** tab, in the **Navigate** group, click **Explore** and then move to a different area in the image. Click the **Segment Picker** tool to activate it so you can collect more training samples.

The bottom section of the **Image Classification** pane shows you all the training samples for each class. You can view and manage training samples by adding, grouping, or removing them. You can remove training samples individually, or you can group them together by selecting them and clicking the **Delete** button. When you select a training sample, it is selected on the map. Double-click a training sample in the table to zoom to it on the map.

5. For each subclass under **Impervious** and **Pervious**, collect about 20 training samples. Make sure that you collect samples that vary a bit in color, shade, and texture to facilitate an accurate classification.

   After collecting the training samples for all classes, you'll need to save them.

6. In the bottom section of the **Image Classification** pane, click **Save As**. Navigate to the **C:\Top20Imagery\Top20Imagery_12\Training** folder and give your training samples file a name, such as CharlottePervImperv. Click **Save**.

Train the classifier

The first step is to use the training samples you collected to train the classifier.

1. In the **Geoprocessing** pane, search for and open the **Train Random Trees Classifier (Image Analyst Tools)** tool and apply the following settings:
   - **Input Raster**: Charlotte_NAIP2023.crf
   - **Input Training Sample File**: CharlottePervImperv
   - **Output Classification Definition File**: CharlottePervImperv.ecd
   - **Additional Input Raster**: SegmentedCharlotteNAIP.tif
   - **Max Number of Trees**: 150
   - **Maximum Tree Depth**: 50
   - **Max Number of Samples Per Class**: 10,000

2. Expand the **Segmented Attributes** section. Check the boxes for the following attributes:
   - **Converged color**
   - **Mean digital number**
   - **Standard deviation**
   - **Rectangularity**

3. Run the tool.

   The training file will take several minutes to be generated.

Classify the image

1. The next step is to classify the image using the training file.

2. Search for and open the **Classify Raster (Image Analyst)** geoprocessing tool and apply the following settings:
   - **Input Raster**: Charlotte_NAIP2023.crf
   - **Input Classifier Definition File**: CharlottePervImperv.ecd
   - **Output Classified Raster**: Charlotte_ClassMap.tif
   - **Additional Input Raster**: SegmentedCharlotteNAIP.tif

3. Run the tool.

> **Note:** If you include an **Additional Input Raster**, it must be the same **Additional Input Raster** used in the **Train Random Trees** tool.

The result, **Charlotte_ClassMap.tif**, is created in the output folder you specified and added to the **Contents** pane.

Reclassify

It is expected that some pixels and objects will be misclassified. Because subclasses will be combined into **Impervious** or **Pervious** parent classes, misclassification may not be an issue. However, if a subclass is misclassified into the incorrect parent class, it should be reclassified into the proper subclass. You will use the **Reclassifier** tools to do this.

1. Examine the classified raster (**Charlotte_ClassMap.tif**) and note any major discrepancies between the **Impervious** and **Pervious** parent classes.

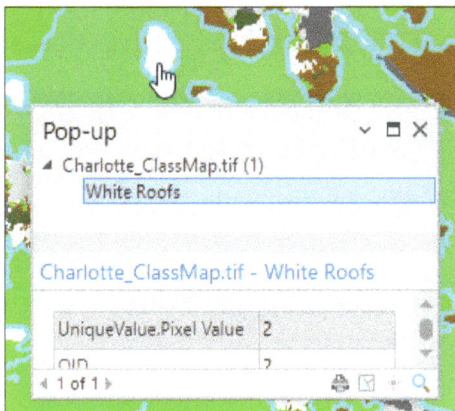

In this example, some of the dormant grass on the golf course in the lower right of the image has been misclassified to **White Roofs**, probably because some sand traps were included in the training sample segment.

2. In the **Contents** pane, make sure **Charlotte_ClassMap** is selected.

3. Click the **Imagery** tab. In the **Image Classification** group, click **Classification Tools** and then click **Reclassifier**.

    The **Reclassifier** pane appears.

4. Under **Edit Type**, click **Reclassify within a region**.

5. Under **Remap Classes**, click **Current Class** and then click **White Roofs** (incorrect class). Then click **New Class** and click **Grass Turf** (correct class).

6. On the map, draw a polygon boundary of the incorrect class to fix it. Double-click the last point to complete the polygon.

    The current class is reclassed to the correct one.

7. At the bottom of the pane, for **Output Dataset**, set the name to Charlotte_ClassMap2.tif and click **Run**.

    The updated class map is created.

Merge classes

After the class map is reclassified and created, you will merge the various subclasses into their parent classes.

1. Make sure the new class map (**Charlotte_ClassMap2.ti**f) is active in the **Contents** pane.

2. On the **Imagery** tab, click **Classification Tools** and then click **Merge Classes**.

    The **Merge Classes** tool appears. All the subclasses you specified and collected training samples for are displayed in a list under the heading **Old Class**.

3. For **Classification** schema, click the **Browse** button and then click **Impervious_Schema.ecs**.

4. In the **New Class** column, click the arrow and select either the **Impervious** or **Pervious** parent class as appropriate.

5. For **Output Dataset**, set the name of the dataset to Charlotte_ClassMapMerge.

Old Class	New Class	
▨ Asphalt	▨ Impervious	⌄
☐ Concrete	▨ Impervious	⌄
☐ White Roofs	▨ Impervious	⌄
▨ Gray Roofs	▨ Impervious	⌄
▨ Grass Turf	▨ Pervious	⌄
▨ Bare Trees	▨ Pervious	⌄
▨ Leafy Trees	▨ Pervious	⌄
▨ Bare	▨ Pervious	⌄

6. Click **Run**.

   The **Charlotte_ClassMapMerge.crf** file is generated and saved to the folder.

Apply data formatting

The **Charlotte_ClassMapMerge** data contains class values, labels, symbols, and other data associated with a class map. The raster ground truth dataset includes only two values, 0 representing impervious surfaces and 1 representing pervious surfaces. Therefore, in this example, you need to convert the class map to match the ground truth.

1. Click the **Analysis** tab. In the **Geoprocessing** group, click **Tools**.

2. In the **Geoprocessing** pane, search for and open the **Reclassify (Spatial Analyst Tools)** tool. Apply the following settings:
   - **Input raster**: Charlotte_ClassMapMerge
   - **Reclass field**: Value

3. Fill out the **Reclassification** table as follows:

Value	New
0	0
1	1
NODATA	NODATA

4. For **Output raster**, save the raster as Charlotte_ClassMapMerge2.

5. Click **Run**.

Conduct accuracy assessment

Accuracy assessment is a crucial part of the image classification workflow because it quantifies the degree of error in the classified map, identifying how well each class represents real-world conditions and highlighting specific areas of misclassification.

1. In the **Geoprocessing** pane, search for and open the **Create Accuracy Assessment Points (Image Analyst Tools)** tool and apply the following settings:
   - **Input Raster or Feature Class Data**: Charlotte_ClassMapMerge2
   - **Output Accuracy Assessment Points**: Charlotte_GroundTruthTest
   - **Target Field**: Ground truth
   - **Number of Random Points**: 150
   - **Sampling Strategy**: Equalized stratified random

2. Run the tool.

3. Search for and open the **Update Accuracy Assessment Points (Image Analyst Tools)** tool and apply the following settings:
   - **Input Raster**: Charlotte_ClassMapMerge2
   - **Input Accuracy Assessment Points**: Charlotte_GroundTruthTest

- **Output Accuracy Assessment Points**: Charlotte_GroundTruthTest_
update
- **Target Field**: Classified

4. Run the tool.

5. Search for and open the **Compute Confusion Matrix (Image Analyst Tools)**
tool and apply the following settings:
   - **Input Accuracy Assessment Points**: Charlotte_GroundTruthTest_
update
   - **Output Confusion Matrix**: Charlotte_ConfusionMatrix

6. Run the tool.

The confusion matrix is computed and added to the **Contents** pane.

7. In the **Contents** pane, under **Standalone Tables**, right-click **Charlotte_
ConfusionMatrix** and click **Open**.

	OBJECTID *	ClassValue	C_0	C_1	Total	U_Accuracy	Kappa
1	1	C_0	66	15	81	0.814815	0
2	2	C_1	9	60	69	0.869565	0
3	3	Total	75	75	150	0	0
4	4	P_Accuracy	0.88	0.8	0	0.84	0
5	5	Kappa	0	0	0	0	0.68

A confusion matrix, also known as an error matrix, evaluates the performance
of the classification model by comparing the model's predictions with the actual
outcomes. It shows how often the model correctly or incorrectly predicted
impervious and pervious features.

In the confusion matrix table, **C_0** is **Impervious** features and **C_1** is **Pervious**
features. Two types of error are calculated: user error (**U-Accuracy**) and producer error (**P_Accuracy**). User error occurs when a pixel is included in a class
where it doesn't belong. Producer error occurs when a pixel that belongs to a class
is not classified as such. In other words:

- User Error: Wrong assignment to a class
- Producer Error: Missed inclusion in the correct class

In the table shown, you see that the user error for impervious and pervious
features is 0.81 and 0.87 respectively, whereas the producer error is 0.88 and 0.8,
respectively. The overall error is 0.84, which is fairly accurate.

Also included is the Kappa statistic, which determines how reliable the classification is. A Kappa value of 0.68 is substantial[1 and 2].

Your accuracy assessment results will vary from the preceding example, which used the training data provided with the data download and improved with the reclassification steps described in this section.

Summary

In this chapter, you learned the art and science of multispectral image classification to produce an impervious surface class map. In a series of tutorials, you performed supervised classification based on objects and features in the imagery, defined a classification schema, and collected training samples representing impervious and pervious features, trained the classifier, classified the image, merged classes, and calculated an accuracy assessment of the resulting class map.

These skills and techniques are effective in producing class maps used for environmental compliance, stormwater fee assessment, hydrologic modeling, flood prediction and management, and identifying and implementing best management practices for removing pollution from runoff before reaching surface waters, such as lakes and streams.

These skills and techniques are an important input for decision support scenarios.

Information at your fingertips

The following links are useful for providing information about concepts, techniques, and best practices for classifying multispectral imagery.

- Understand image segmentation and the classification process in ArcGIS Pro: links.esri.com/AccuracySegmentationClassifyProcess
- Understand how to set spectral and spatial detail parameters in the Segment Mean Shift geoprocessing tool: links.esri.com/SegmentationParameters
- Description of the Training Samples Manager, with supporting workflows: links.esri.com/AccuracyTrainingSamplesManager
- Overview of accuracy assessment of classified imagery in ArcGIS Pro: links.esri.com/AccuracyAssessment
- An overview of the Classification and Pattern Recognition toolset: links.esri.com/ClassificationToolset

Notes

1. Landis, J. R., and G. G. Koch. 1977. "The Measurement of Observer Agreement for Categorical Data." *Biometrics* 33 (1): 159–74. https://doi.org/10.2307/2529310.
2. Congalton, Russell G., and Kass Green. 2009. *Assessing the Accuracy of Remotely Sensed Data: Principles and Practices*, 2nd ed. CRC Press.

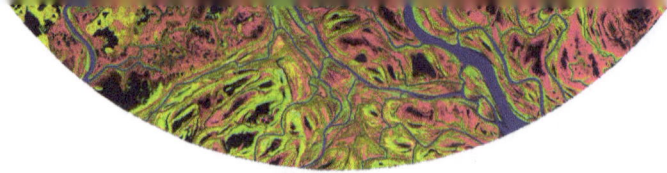

CHAPTER 13
Using deep learning for object detection and classification

Pavan Yadav

Objectives

- Set up ArcGIS Pro for deep learning workflows.
- Create representative samples of the objects of interest to prepare training data.
- Use the training data to train a deep learning model.
- Use the trained deep learning model to perform inference on new, unseen imagery.
- Visualize the result.

Introduction

GeoAI combines artificial intelligence (AI) and geospatial data to analyze and interpret imagery. You will use deep learning methods to extract complex features from images. The first step is to set up your computer by downloading deep learning frameworks and then configuring your machine.

The deep learning workflow typically encompasses the following key stages:

1. **Prepare training data**: Generate training samples of features or objects of interest for the machine to learn.
2. **Train the model**: Use these training samples to train a deep learning model.

3. **Perform inferencing**: Use the trained deep learning model to perform inferencing on new data.

Tutorial 13-1: Explore a deep learning workflow

In this tutorial, you will focus on object detection and instance segmentation, along with classification using the Mask R-CNN deep learning model architecture. You'll tackle a real-world challenge in wildlife management: automating the detection and classification of caribou (both adults and calves) within aerial imagery of Alaskan herds.

Set up ArcGIS Pro for deep learning

ArcGIS Pro deep learning tools use Python deep learning libraries that need to be installed and verified.

1. Make sure that the ArcGIS Pro application is closed.

2. Go to the Deep Learning Libraries Installers for ArcGIS page at links.esri.com /DLFrameworkInstaller.

3. Under **Download**, click **Deep Learning Libraries Installer for ArcGIS Pro 3.5** to download it.

4. In your files, extract **ArcGIS_Pro_35_Deep_Learning_Libraries.zip**.

5. Open the resulting **ArcGIS_Pro_35_Deep_Learning_Libraries** folder.

6. Double-click the **ProDeepLearning.msi** Windows installer package to install the deep learning libraries.

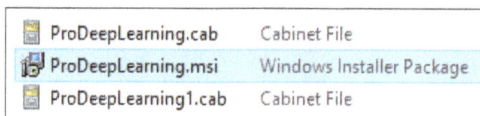

ProDeepLearning.cab	Cabinet File	
ProDeepLearning.msi	Windows Installer Package	
ProDeepLearning1.cab	Cabinet File	

7. Complete the installation process.

Verify that the installation is successful

1. Start ArcGIS Pro.

2. On the left, click **Settings**.

3. Click the **Package Manager** tab.

4. In the search field, type deep. In the results, check that the **deep-learning-essentials** package is listed.

   You installed the deep learning libraries and confirmed their installation in ArcGIS Pro. You are now ready to perform the deep learning analyses.

Download the tutorial data and set up the project

1. Go to links.esri.com/Imagery20Data and download the data for chapter 6.

2. Unzip the folder to **C:\Top20Imagery**.

   > **Note:** In the second chapter, you created a folder named **Top20Imagery** on your C: drive. If you haven't done that, create that folder now. Now and in subsequent chapters, you will download and unzip the data for each chapter to this folder.

3. Inside the **Top20Imagery_13** folder, double-click **Top20Imagery_13.aprx** to open the ArcGIS Pro project for this chapter.

   The map displays a remote area within Noatak National Preserve, Alaska. The **Contents** pane lists two image layers, **Imagery1** and **Imagery2**. The first image shows a small part of a caribou herd. This imagery will be used for generating the training data needed to build a deep learning model for detecting caribou. Once trained, this model will be applied to the second image.

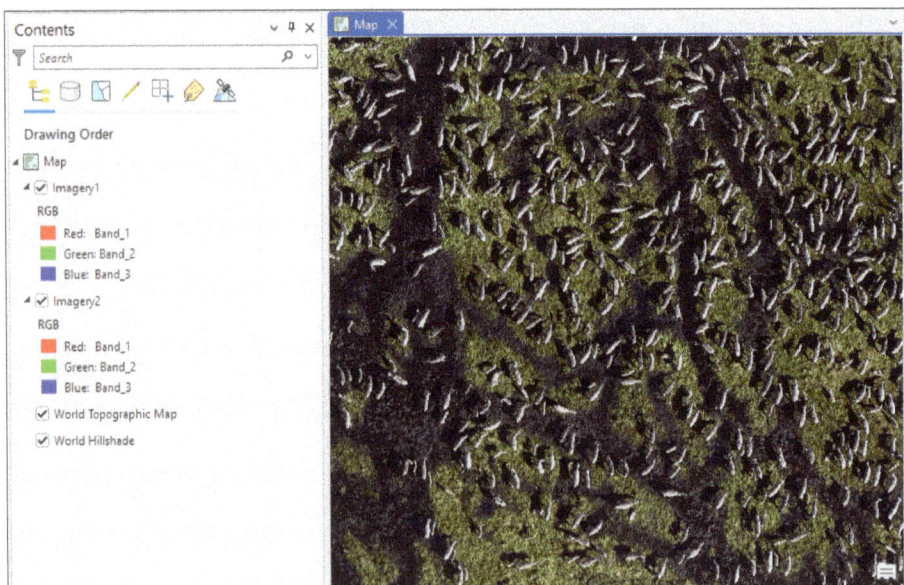

4. In the **Contents** pane, uncheck the box for **Imagery1** to turn off its visibility.

5. View the **Imagery2** layer showing the entire caribou herd. Zoom in and pan to examine the imagery. Observe the adult and calf caribou throughout the landscape.

Prepare your training data

The first step in training the deep learning model is creating training samples. Creating good training samples is essential when training a deep learning model. It is also the most time-consuming step in the process. To provide your deep learning model with the information it needs to extract all the adult and calf caribou in the image, you'll create samples for adult and calf caribou to train the model about the size, shape, and spectral signature of adult and calf caribou. These training samples are created and managed through the **Label Objects for Deep Learning** tool.

> **Note:** Creating a training dataset by digitizing hundreds of features can be time-consuming. To learn how to create the samples, you can create schema and collect all or some samples. If you prefer to skip this step and proceed to the "Create Image Chips" section, use the premade dataset in the **Results** geodatabase within the **Provided Results** folder.

Create the training schema

A classification schema determines the number and types of classes to use. In your case, you will create training data for two classes: Adult and Calf.

1. In the **Contents** pane, turn off **Imagery2** and turn on **Imagery1**. Right-click **Imagery1** and click **Zoom To Layer**.

2. On the ribbon, click the **Imagery** tab. In the **Image Classification** group, click **Deep Learning Tools** and then click **Label Objects for Deep Learning**.

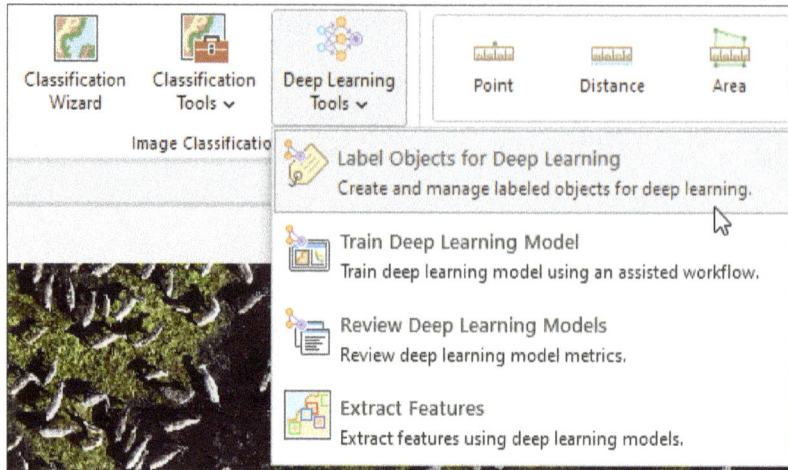

The **Label Objects for Deep Learning** dialog box appears with **Label using** set to the **Existing** imagery layer and the **Imagery layer** set to **Imagery1**.

3. Click **OK**.

   The **Image Classification** pane appears with a blank schema. You'll create a schema with two classes: **Adult** and **Calf**.

4. In the **Image Classification** pane, right-click **New Schema** and click **Edit Properties**.

5. For **Name**, type Caribou. Optionally, set **Organization** name and **Description** for the schema.

6. Click **Save**.

   The schema is renamed in the **Image Classification** pane. You can now add classes to it.

7. Right-click **Caribou** and click **Add New Class**.

8. In the **Add New Class** pane, apply the following settings:
   - **Name**: Adult
   - **Value**: 1
   - **Color**: Mars Red (column 2, row 3)

9. Click **OK**.

10. Repeat steps 6 to 9 to create another class for calves. For this class, set the following:

- **Name**: Calf
- **Value**: 2
- **Color**: Cretan Blue (column 10, row 3)

Now your schema should have the two classes: **Adult** and **Calf**.

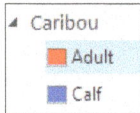

The **Adult** and **Calf** classes are added to the **Caribou** schema in the **Image Classification** pane. You'll create features with these classes to train the deep learning model.

Create training samples

To make sure you're capturing a representative sample of both adult and calf caribou, you'll digitize features throughout the image. These features are read into the deep learning model in a specific format called image chips. Image chips are small blocks of imagery extracted from the source image containing labeled training samples. Once you've created enough training samples in the **Image Classification** pane, you'll export them as image chips with their corresponding labels using the **Export Training Data for Deep Learning** tool.

1. On the ribbon, click the **Map** tab. In the **Navigate** group, click **Bookmarks** and select **Training Location1**.

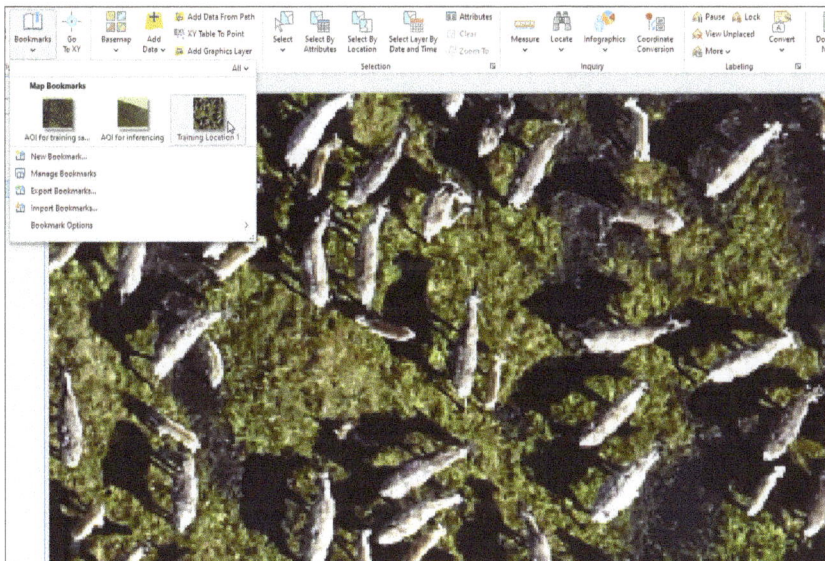

The map zooms to the area of sample caribou that you'll identify.

2. In the **Image Classification** pane, click the **Adult** class. Above the **Caribou** class heading, click the **Rectangle** tool.

    You'll use this tool to draw a rectangle around each adult caribou in your current display. Each rectangle is created by three points.

3. To draw a rectangle around a single caribou, on the map, click to create the first corner, move the pointer, and click the map to specify the direction of the feature. Move the pointer to specify the length and width and click the map to create the rectangle.

A new adult caribou record is added to the **Labeled Objects** group of the **Image Classification** pane.

4. Draw rectangles around each adult caribou in the map display.

5. To create a sample for calves, in the **Image Classification** pane, click the **Calf** class and then click the **Rectangle** tool and repeat the previous steps.

6. Create a record for every caribou, adult or calf, in the image.

    It's essential to label all caribou. If you leave some unlabeled, the model won't know if they're caribou or a different object during training, making the training process harder and less efficient.

> **Tip:** If you need extra guidance to understand how to draw these rectangles or if you want to skip the labor-intensive process of digitizing caribou, a training sample dataset is available in the folder you downloaded. On the ribbon, on the **Map** tab, in the **Layer** group, click **Add Data**. Browse to the **Databases** folder and double-click the **Results** geodatabase. Click **CaribouSamples** and click **OK**.
>
> To distinguish adults and calves, change the symbology. Make sure the **Caribou-Samples** layer is selected in the **Contents** pane, click the **Feature Layer** tab on the **Symbology** pane, and under **Primary symbology**, click **Unique values > Classname** in the **Field 1** list.

When you're finished labeling this image, you'll have approximately 600 samples recorded in the **Training Samples Manager** pane.

Here are a few details to help you as you label the image:

- You can zoom and pan around the map to make digitizing easier.
- It is OK if the rectangles you draw overlap.
- Your final model will account for the size of the adult and calf caribou you identify, so be sure to label all adults and calves as accurately as possible.

7. When you're finished creating samples, in the **Image Classification** pane, click **Save**.

8. In the **Save current training samples** window, under **Project**, click **Databases** and double-click the default project geodatabase, **Caribou.gdb**.

9. Name the feature class CaribouSamples and click **Save**.

10. Close the **Image Classification** pane. If the **Label Objects** window appears, click **Yes**.

11. In the **Catalog** pane, expand the **Databases** folder. Right-click **Caribou.gdb** and click **Refresh**.

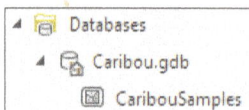

The **CaribouSamples** feature class is now visible.

12. On the **Quick Access** toolbar, click **Save**.

Create image chips

1. On the **Analysis** tab, in the **Geoprocessing** group, click **Tools**.

2. In the **Geoprocessing** pane, search for and open the **Export Training Data for Deep Learning (Image Analyst Tools)** tool. Apply the following settings:
   - **Input Raster**: Imagery1
   - **Output Folder**: imagechips
   - **Input Feature Class Or Classified Raster Or Table**: Caribou.gdb > CaribouSamples
   - **Class Value Field**: Classvalue
   - **Metadata Format**: R-CNN Masks

3. At the top of the tool pane, click the **Environments** tab.

4. Under **Raster Analysis**, for **Cell Size**, click **Same as layer Imagery1**.

5. Click **Run**.

   Note: Depending on your computer's hardware, the tool will take a few minutes to run.

   The images chips are created and ready to be used for training a deep learning model.

6. Save your project.

   You have successfully created training samples and exported them to a format compatible with a deep learning model for training.

Train a deep learning model

You will use the **Train Deep Learning Model** geoprocessing tool to train the model. The tool uses the exported image chips with their corresponding labels to determine what combinations of pixels in the image represent adult and calf caribou. You'll use these training samples to train a Mask R-CNN deep learning model.

Depending on your computer's hardware, training the model can take about an hour. It's recommended that your computer be equipped with a dedicated GPU. If you do not want to train the model, a deep learning model has been provided in the project's **Provided Results** folder and you can skip ahead to the "Detect Caribou" section.

1. Search for and open the **Train Deep Learning Model** tool. Apply the following settings:
   - **Input Training Data**: imagechips
   - **Output Model**: Model_Caribou_Detection
   - **Max Epochs**: 50
   - **Model Type**: MaskRCNN (Object detection)

2. Expand the **Data Preparation** section of the tool.

   The default **Batch Size** for the Mask R-CNN is 4. On a machine with 8 GB GPU, a batch size of 4, and 50 epochs, the tool takes about 25–30 minutes. You may increase or decrease the **Batch Size** per your GPU.

3. Click **Run**.

   The tool prepares the data, performs data augmentation, and sets the correct internal parameters to create a good model. Normalization and augmentation, such as contrast, brightness, and rotation, are automatically performed as part of the training process.

   > **Note:** If the model fails to run, reducing the **Batch Size** parameter can help. You may have to set this parameter to 2 or 1 and rerun the tool.

Detect caribou

You'll use the trained model to detect caribou throughout the target imagery, which is called inferencing.

> **Note:** If you did not train a model in the previous section, a deep learning trained model package has been provided for you in the **Results** folder.
>
> Deep learning inferencing can be a GPU-intensive process and can take time to complete depending on your computer's hardware. If you choose not to detect the caribou on your machine, results have been provided.
>
> In the **Contents** pane, uncheck the **Imagery1** layer. Check the **Imagery2** layer, right-click, and then click **Zoom To Layer**. This image is for the entire area of interest for which you are going to detect caribou.

1. Search for and open the **Detect Objects Using Deep Learning** tool and apply the following settings:

   - **Input Raster**: Imagery2
   - **Output Detected Objects**: DetectedCaribou
   - **Model Definition**: Model_Caribou_Detection > Model_Caribou_Detection.emd
   - **Batch Size**: 8
   - **Confidence Threshold**: 0.5

   Note: If you did not train a deep learning model, browse to the project folder. Open **Provided Results**. Open **Model_Caribou_Detection** and click the **Model_Caribou_Detection.emd**. Click **OK**.

2. Click the **Environments** tab.

3. Under **Raster Analysis**, for **Cell Size**, click **Same as layer Imagery2**.

4. Click **Run**.

   The tool will take several minutes to run, depending on your hardware and whether you are running on CPU, GPU, or RAM.

   Observe your results. You can try experimenting with the arguments to see how this impacts your results.

5. In the **Detect Objects Using Deep Learning** tool, on the **Environments** tab, for **Processing Extent**, click **Default**.

6. Click **Run**.

   Note: If you do not run the model to detect and classify the caribou, a dataset of caribou detected and classified has been provided. To add the feature class to the map, on the ribbon, on the **Map** tab, in the **Layer** group, click **Add Data**. Browse to the **Caribou > Provided Results** folder, open the **Results** geodatabase, and double-click the **DetectedCaribou** feature class.

7. When the tool finishes, observe your results. The color of your results may differ from the image provided.

8. In the **Contents** pane, turn on the **Imagery2** layer and turn off the **Caribou-Samples** layer.

Prepare the detected features for display

1. In the **Contents** pane, select the **DetectedCaribou** feature layer.

2. On the **Feature Layer** tab, in the **Drawing** group, click **Symbology**.

3. In the **Symbology** pane, change the **Primary symbology** to **Unique Values**.

4. For **Field 1**, click **Class**.

5. On the **Classes** subtab, click the **Add all values** button.

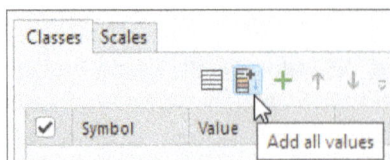

6. Click the first symbol, which is for **Adult**.

    The **Format Polygon Symbol** pane appears. Make sure you are on the
    Properties tab.

7. Under **Appearance**, apply the following settings:
    - **Color**: No color
    - **Outline color**: Mars Red
    - **Outline width**: 1 pt

8. Click **Apply**.

9. Click the back button.

10. Click the second symbol, which is for **Calf**.

11. Under **Appearance**, apply the following settings:
    - **Color**: No color
    - **Outline color**: Cretan Blue
    - **Outline width**: 1

12. Click **Apply**.

13. Click the back button.

14. In the **Classes** table, right-click **<all other values>** and click **Remove**.

    You have now configured the symbology for the **DetectedCaribou** layer.

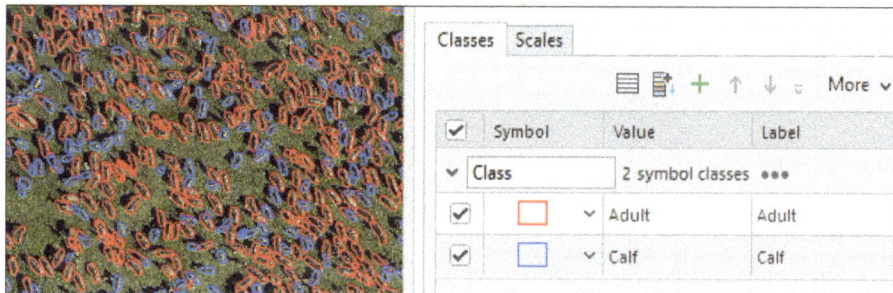

15. Close the **Symbology** pane.

Remove outliers

You will remove detections that are unusually large or small. You can use either the perimeter or the area of the detected shapes to do this. You'll eliminate any detections with a perimeter smaller than 2 meters.

1. In the **Contents** pane, right-click the **DetectedCaribou** layer and click **Attribute Table**.

2. On the table toolbar, click **Select By Attributes**.

3. Create the following query expression:
   - **Where**: Shape_Length
   - **Operator**: is less than
   - **Value**: 2.0

| Where | Shape_Length | ⌄ | is less than | ⌄ | 2.0 | ⌄ | ✕ |

4. Click the **Verify** button (check mark) to validate the expression.

5. Click **OK**.

   The attributes that match the expression are selected on the map and in the table.

6. On the table toolbar, click the **Delete** button.

   The selected features are deleted.

7. Close the attribute table.

8. Save your project.

Summary

In this chapter, you learned how to detect features in imagery using deep learning techniques. You learned how to properly set up your computer to support deep learning workflows, collect and label training samples, create the deep learning model, and perform inferencing (classification). Lastly, you visualized the results and removed discrepancies.

Information at your fingertips

Introduction to deep learning

For more information about how deep learning works to detect and classify objects in imagery, see links.esri.com/how-DL-works.

Deep learning content

The Esri Analytics team provides an extensive collection of ready-to-use deep learning packages that can be used across the ArcGIS system. Deep learning packages developed using specific types of imagery and deep learning frameworks may not produce similar results using imagery that has different spectral and spatial characteristics. See links.esri.com/Deep_Learning_Content.

Install the deep learning frameworks tutorial

For more information about installing deep learning frameworks in ArcGIS, see links.esri.com/Install_DL_frameworks_tutorial.

How Mask R-CNN works

For more information about how the Mask R-CNN deep learning framework used in this chapter works, see links.esri.com/how-maskrcnn-works.

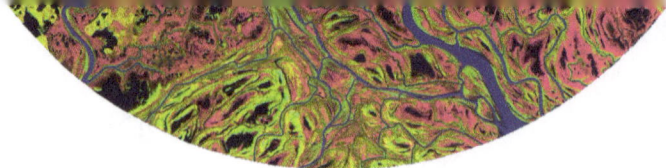

CHAPTER 14
Visualizing and analyzing scientific multidimensional raster data

Hong Xu

Objectives

- Prepare a multidimensional raster for climate data analysis.
- Describe how to display and retrieve slices of a multidimensional raster for a given scenario.
- Prepare the time series data of a multidimensional raster for a trend analysis.
- Apply a trend analysis to a multidimensional raster for a given scenario.

Introduction

In earth science, scientific data related to weather, the oceans, the atmosphere, and land can be measured at specific dates and times. These measurements are recorded as additional dimensions as part of the collected information. These measurements, referred to as variables, retain information related to the time of collection and the type of data collected. They can be data points, such as temperature, humidity, wind speed, salinity, or some other measured quantity. When using ocean or atmospheric data, you can include depths or altitudes. When data of this type is collected—data with a regular spatial component combined with a time component and additional information for recorded variables—it is called multidimensional data. ArcGIS Pro imports this data as rasters, and you can use specific techniques to understand and analyze these multidimensional rasters.

Multidimensional rasters allow you to explore trends and spatial pattens in the data. A variable represents the scientific measurements across space and time. For example, when examining a precipitation variable, you may want to understand how precipitation changes over several years.

Tutorial 14-1: Add a multidimensional raster to a map and explore it

In this tutorial, you will learn how to add a multidimensional raster file to a map in ArcGIS Pro and use interactive tools to explore the data. These include visualizing slice by slice (2D) using the **Multidimensional** tab and exploring a time series (1D) using a temporal profile.

Download the tutorial data and set up the project

1. Go to links.esri.com/Imagery20Data and download the data for chapter 14.

2. Unzip the folder to **C:\Top20Imagery**.

> **Note:** In the second chapter, you created a folder named **Top20Imagery** on your C: drive. If you haven't done that, create that folder now. Now and in subsequent chapters, you will download and unzip the data for each chapter to this folder.

3. Inside the **Top20Imagery_14** folder, double-click **Top20Imagery_14.aprx** to open the ArcGIS Pro project for this chapter.

Add the multidimensional imagery to a map and display it

1. On the ribbon, on the **Map** tab, click the **Add Data** arrow and then click **Multidimensional Data**.

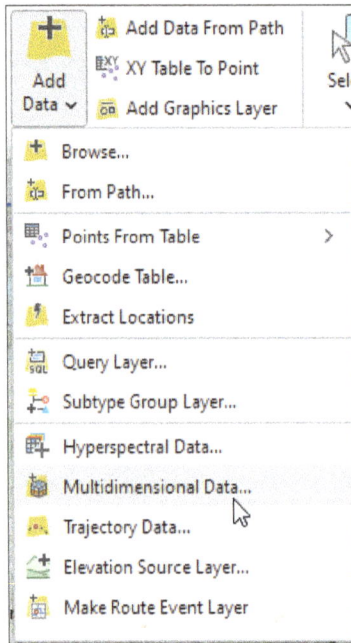

2. In the **Add Multidimensional Data** dialog box, click the **Browse** folder icon.

3. Browse to **C:\Top20Imagery\Top20Imagery_14** and click **precip.V1.0.mon. mean.nc**.

   Variables in the NetCDF file show in the **Select Variable** list. This multidimensional raster has only one variable, the daily precipitation amounts, represented as a monthly average.

4. Check the box next to **precip** to select the variable and click **OK**.

   > **Note:** NetCDF is a file format for organizing, storing, processing, and sharing arrays of scientific data. It's used in many fields, including atmospheric and oceanographic science.

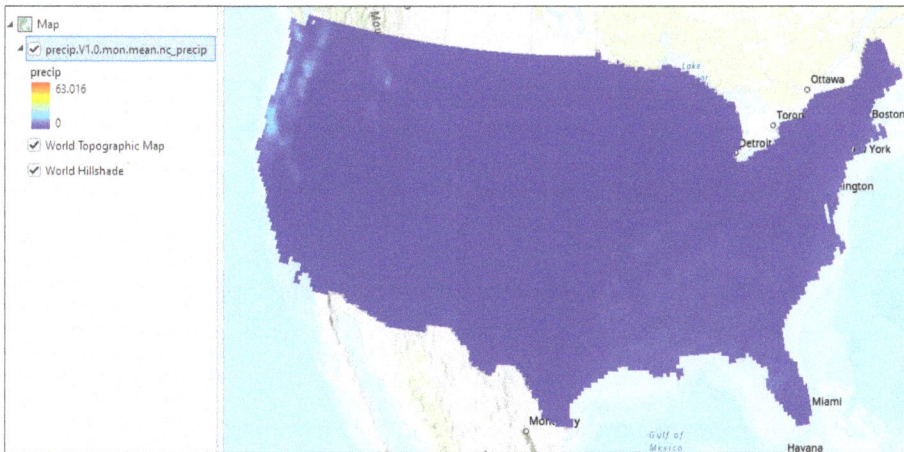

The layer is added to the map. As with other layers in ArcGIS Pro, the multidimensional layer activates specific contextual tabs; in addition to a **Raster Layer** and a **Data** tab, you also have access to a **Multidimensional** tab. This tab provides access to visualization, exploration, data management, and specific tools for the analysis of multidimensional rasters.

Because layer statistics have not been calculated for the data, you can enhance the display by dynamically computing statistics.

5. Click the **Raster Layer** tab. In the **Rendering** group, click **DRA**.

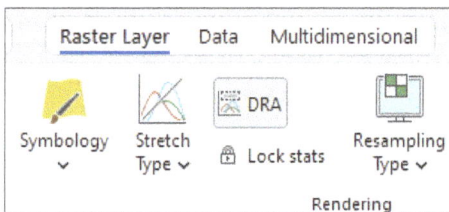

The display in the map is enhanced.

Explore a multidimensional raster slice by slice

Next, you will visualize measured precipitation data in a time series array from January 1948 to May 2024. Then you will plot this data in a graph to analyze trends.

1. Click the **Multidimensional** tab.

    In the **Current Display Slice** group, the earliest date is displayed by default. In this instance, the first variable displayed is for precipitation recorded in January 1948 (1948-01-01). You can select and view data from any of the recorded, discrete time stamps.

2. In the **Current Display Slice** group, click **StdTime** and click **precip**.

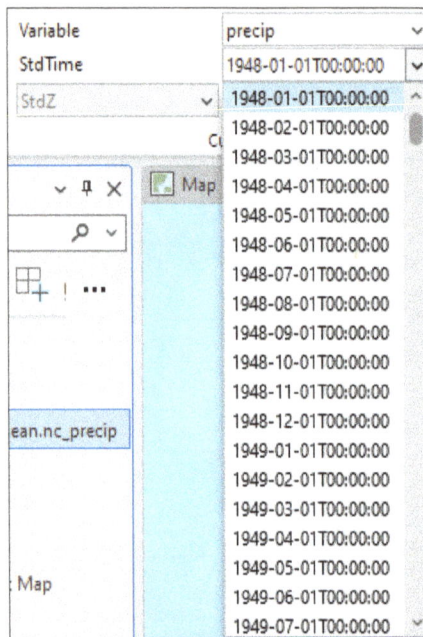

    You can scroll through the list to see the recorded variables of measured precipitation available in the time array from January 1948 to May 2024.

3. Find and select the **2024-01-01T00:00:00** time slice to visualize that month's data in the map.

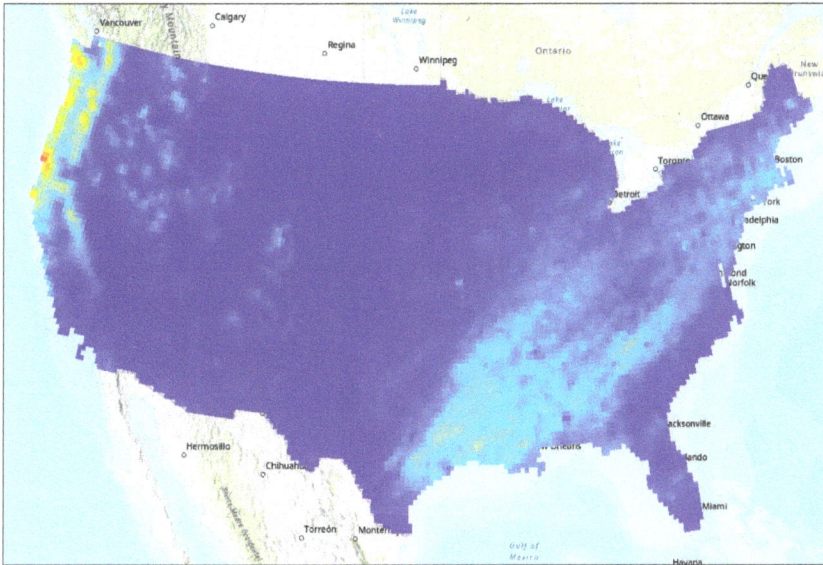

4. In the **Current Display Slice** group, click the left or right arrow to display the slices month by month.

5. Click the play button to cycle through the data slices in an animation.

   The specific slice of time will be listed in the **StdTime** box as it is displayed in the map.

6. Stop the animation when you are finished and then set the current time slice to **1948-01-01T00:00:00**.

Tutorial 14-2: Extract and visualize a precipitation time series

In this tutorial, you will use a multidimensional raster to explore the precipitation change over time using the **Temporal Profile** tool.

1. In the **Contents** pane, make sure that the layer **precip.V1.0.mon.mean.nc_ precip** is selected.

2. On the **Multidimensional** tab, in the **Analysis** group, click the **Temporal Profile** tool.

   The chart window and the **Chart Properties** pane appear.

3. In the **Chart Properties** pane, under **Define an area of interest**, click the **Point** tool.

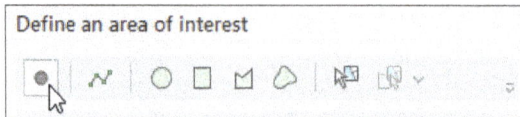

4. On the map, click a location in the Pacific Northwest.

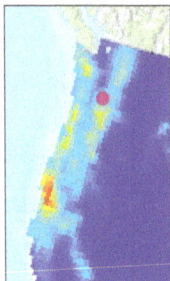

5. In the **Chart Properties** pane, under **Symbol**, click the point to change the color to an obvious color.

6. At the top of the **Chart Properties** pane, click the **General** tab. Update the following settings:
   - **Chart title**: Change in monthly precipitation over StdTime
   - **Y-axis title**: Accumulated Precipitation

7. On the chart toolbar, click the **Legend** button to turn it off and make more space for the chart.

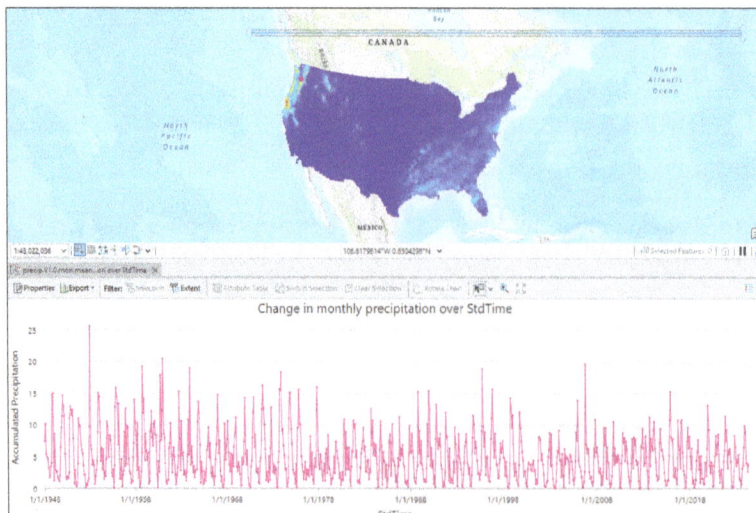

Your display may look slightly different depending on where you clicked in the map. You now have a temporal profile chart of accumulated monthly precipitation for a single, defined location.

When you're observing data for such a long period, any pattern may not be obvious due to the seasonal variation and dense observations.

8. In the **Chart Properties** pane, return to the **Data** tab. Click the arrow next to **Aggregation Options** to expand the section.

9. For **Time binning options**, change the **Interval size** by clicking the calendar and update to **1 Years**.

10. Expand the **Trend** section. Check the box for **Show Trend Line**.

The updated chart now shows a trend line.

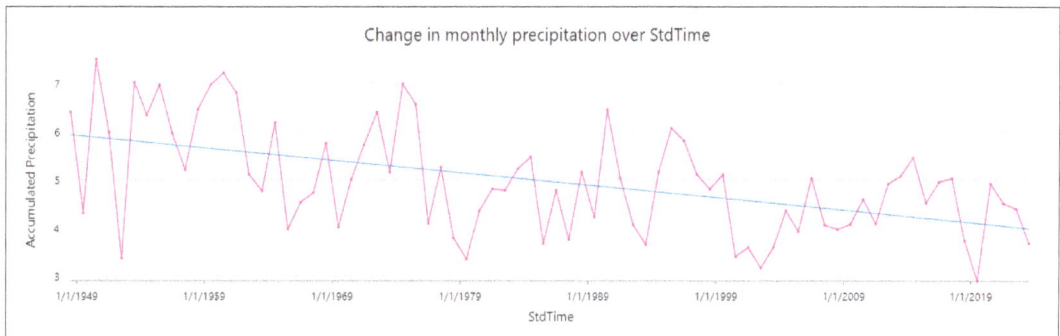

The chart shows the annual precipitation profile and an overall decreasing trend for your selected location.

11. Uncheck the first box for **Location 1** to turn off the profile for that location.

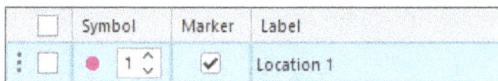

Analyze the precipitation trend of two states using multidimensional analysis tools

Next, you will analyze the precipitation trend of two states , Washington and New York. You can simply draw polygons interactively or use state boundaries from ArcGIS Online.

1. In the **Catalog** pane, click the **Portal** tab and then click the **Living Atlas** icon.

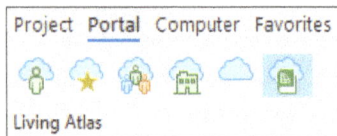

2. In the search field, type USA States Generalized Boundaries.

3. In the results, right-click **USA States Generalized Boundaries** and click **Add To Current Map**.

4. In the **Contents** pane, under the **USA States Generalized** layer, click the rectangle to open the **Symbology** pane.

5. In the **Symbology** pane, on the **Gallery** tab, click **Black Outlined (2 pts)**.

6. Return to the **Chart Properties** pane for the **precip.V1.0.mon.mean.nc_precip** layer. Under **Define an area of interest**, click the **Feature Selector** tool.

7. In the map, click the states of New York and Washington to compare them.

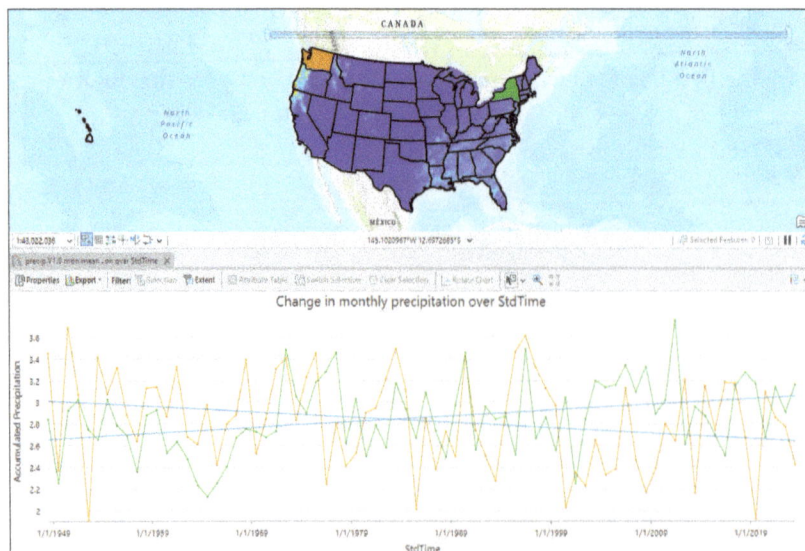

Precipitation for Washington state exhibits more variability than New York state, and precipitation over time is decreasing for Washington and increasing for New York.

8. Close the chart window.

Use multidimensional analysis tools to analyze soil moisture correlation

Next, you will use multidimensional analysis tools to map trends and anomalies of precipitation over time and calculate the correlation with an accompanying soil moisture variable. You will use the Seasonal-Kendall method to map precipitation trends from the monthly precipitation.

Note: The Seasonal-Kendall trend type is a nonparametric method that analyzes data for monotonic trends in seasonal data. The output from the analysis is presented as Sen's slope, which indicates the distribution of the trend. This approach involves computing slopes for all the pairs of ordinal time points and then using the median of these slopes as an estimate of the overall slope. It is expressed as the change in the response variable per unit change in time.

1. In the **Contents** pane, uncheck the box next to **USA States Generalized** to hide the layer. Then, select the **precip.V1.0.mon.mean.nc_precip** layer.

2. On the **Multidimensional** tab, in the **Analysis** group, click the **Trend** tool.

   The **Geoprocessing** pane appears with the settings for the **Generate Trend Raster** tool.

3. In the **Generate Trend Raster** geoprocessing pane, apply the following settings:
   - **Input Multidimensional Raster**: precip.V1.0.mon.mean.nc_precip
   - **Output Multidimensional Raster**: precip_Trend.crf
   - **Trend Type**: Seasonal-Kendall
   - **Seasonal Period**: Months
   - Make sure the check box for **Ignore NoData** is checked.

| Input Multidimensional Raster |
| precip.V1.0.mon.mean.nc_precip |

Output Multidimensional Raster

precip_Trend.crf

Dimension

StdTime

Variables [Dimension Info] (Description)   Select All ↺

☑ precip [StdTime=917] (Monthly Average of Daily Accumul...

Trend Type

Seasonal-Kendall

Seasonal Period

Months

☑ Ignore NoData

4. Click **Run**.

The output layer from the Seasonal-Kendall analysis is added to the map.

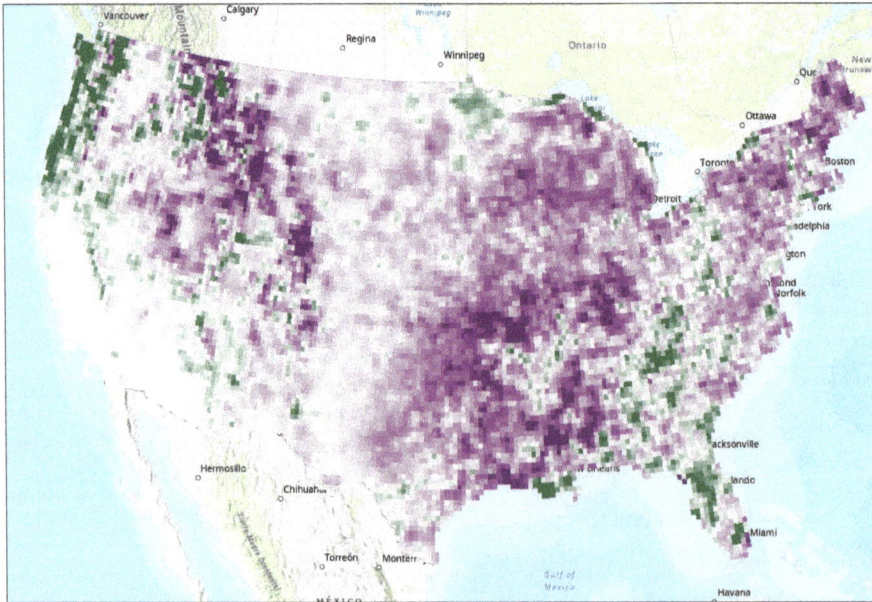

Band 1 of the output raster is automatically displayed in the map with the **Multi-part Color Scheme** symbology. Band 1 represents the Sen's slope, which depicts the distribution of the trend. Purple indicates an increasing trend of precipitation, green indicates a decreasing trend, and white means no change. As we noted in the plots and trend lines in the previous section, precipitation decreases over time in Washington state and increases in New York state.

Take the next step

Multidimensional data represents data captured at multiple times or at multiple depths or heights. This data type is commonly used in atmospheric, oceanographic, and earth sciences. Multidimensional raster data can be captured by satellite observations, in which data is collected at certain time intervals, or generated from numerical models, in which data is aggregated, interpolated, or simulated from other data sources.

Another term for a multidimensional raster is an image cube. An image cube contains one or multiple variables, where each variable is a 3D cube with dimensions of time, latitude, and longitude or a 4D cube with an additional depth dimension included.

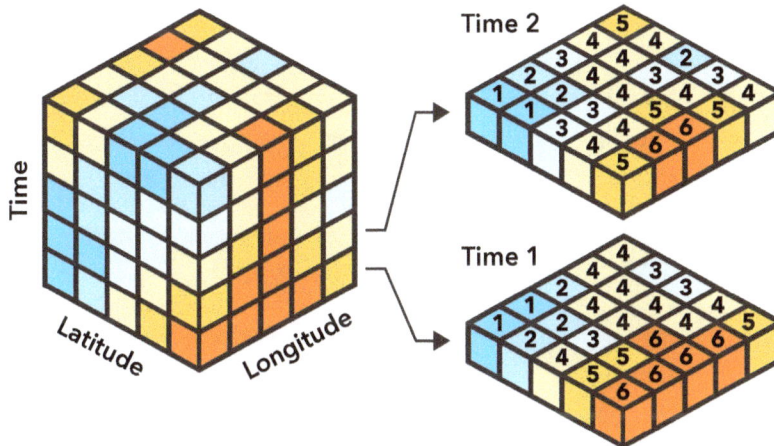

The formats capable of storing multidimensional data include NetCDF, GRIB, and HDF. Additionally, cloud-optimized formats, such as Cloud Raster Format (CRF) and Zarr, have also been developed to store this type of data. ArcGIS supports all these multidimensional formats, enabling you to visualize, manage, and analyze these multidimensional rasters.

Working with multidimensional raster data can be challenging, especially when you have many slices of data and multiple variables. Using the Multidimensional tab, you can visualize the data and perform analysis.

Work with the Multidimensional tab

The Multidimensional tab is contextual, and enabled when you add a multidimensional dataset to a map and select it in the Contents pane. When the Multidimensional tab is enabled for a multidimensional raster layer or a multidimensional mosaic layer, it allows you to manipulate the spatial and dimensional data, display the slice in the map view, open the temporal profile chart, and perform data management and analysis.

Multidimensional geoprocessing tools and raster functions

A suite of multidimensional geoprocessing tools is available in the Multidimensional Analysis toolset, available with the ArcGIS Image Analyst extension. You can aggregate a data cube, find anomalies, and analyze trends by performing analysis across dimensions and for multiple variables.

In addition to analysis, the tools in the Multidimension toolbox allow you to manage multidimensional data, and they do not require an Image Analyst or ArcGIS Spatial Analyst™ license. Many of these tools are also available as raster functions.

The Generate Trend Raster tool is used to analyze trends and supports several methods.

Linear and Hamonic methods are mathematical fitting models that model linear trends and seasonal characteristics, such as trends with periodic fluctuations. The output stores model coefficients as bands, with the slope band often used for mapping trends. These models can also predict data for time periods not included in the input cube. Mann-Kendall and Seasonal-Kendall are nonparametric statistical methods for detecting trends. In these methods, the slope band in the output represents **Sen's Slope**, a median slope calculated from all time pairs. This measure is particularly effective for analyzing trends in climate variables.

Summary

In this chapter, you added a multidimensional raster to a map and explored ways to visualize and interact with it using various tools. You also created a multidimensional trend analysis chart to understand the data over a multidecade span. These skills will equip you to effectively analyze climate data.

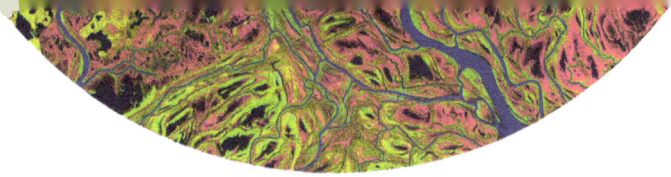

CHAPTER 15
Exploring hyperspectral imagery and spectral analysis

Hong Xu

Objectives

- Evaluate hyperspectral imagery for a given scenario.
- Recognize ways to explore hyperspectral imagery for analysis.
- Classify features using hyperspectral imagery and analysis methods for a given scenario.

Introduction

Hyperspectral imagery, also known as imaging spectroscopy, is imagery collected by a sensor across a wide range of the electromagnetic spectrum in many—usually several hundred—narrow, contiguous bands. This high spectral resolution enables detection and identification of materials that are not possible in regular multiband images. Hyperspectral imagery is often used in mineral exploration, environmental monitoring, precision agriculture, military surveillance, and more.

There are many types of hyperspectral imagery collected by various sensors and platforms: Hyperion (EO-1), AVIRIS (Airborne Visible/Infrared Imaging Spectrometer), EMIT (Earth Surface Mineral Dust Source Investigation), and Wyvern, to name a few. Many of them are public domain data, which can be accessed on the web, and some are commercial data, such as Wyvern. In this chapter, you'll use airborne

AVIRIS imagery to learn the characteristics of hyperspectral imagery and use these characteristics to identify and isolate oak tree locations in a nature preserve.

Tutorial 15-1: Explore hyperspectral imagery

First, you'll sign in to ArcGIS Online to download data used in the tutorial. Next, you'll add the data to ArcGIS Pro using the **Add Hyperspectral Data** dialog box. You'll then explore the band composition using the wavelength-based selection method and image spectra using the spectral profile chart.

Download the tutorial data and set up the project

1. Go to links.esri.com/Imagery20Data and download the data for chapter 15.

2. Unzip the folder to **C:\Top20Imagery**.

> **Note:** In the second chapter, you created a folder named **Top20Imagery** on your C: drive. If you haven't done that, create that folder now. Now and in subsequent chapters, you will download and unzip the data for each chapter to this folder.

3. Inside the **Top20Imagery_15** folder, double-click **Top20Imagery_15.aprx** to open the ArcGIS Pro project for this chapter.

Add hyperspectral imagery

Imagery providers may use different formats to deliver hyperspectral imagery. The **Add Hyperspectral Data** dialog box allows you to work with these formats easily. In this section, you'll add the AVIRIS data to your project.

1. On the ribbon, click the **Map** tab. In the **Layer** group, click **Add Data** and then click **Hyperspectral Data**.

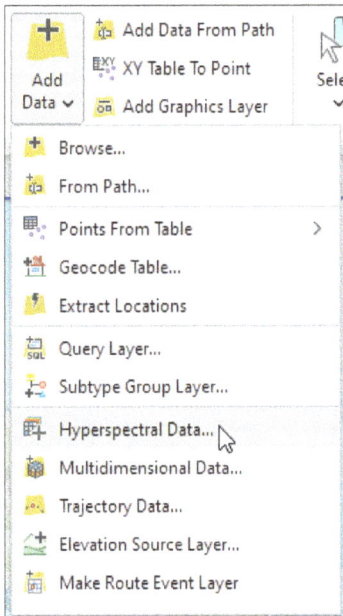

2. In the **Add Hyperspectral Data** window, for **Input Image File**, click the **Browse** button. Navigate to **C:\Top20Imagery\Top20Imagery_15** and click **ang20220420t20401_JL_Preserve.crf**.

3. For **Band Selection Method**, make sure **All** is selected.

4. Click **OK** to add the image data to the map.

The image data displays the Jack and Laura Dangermond Preserve at Point Conception, California.

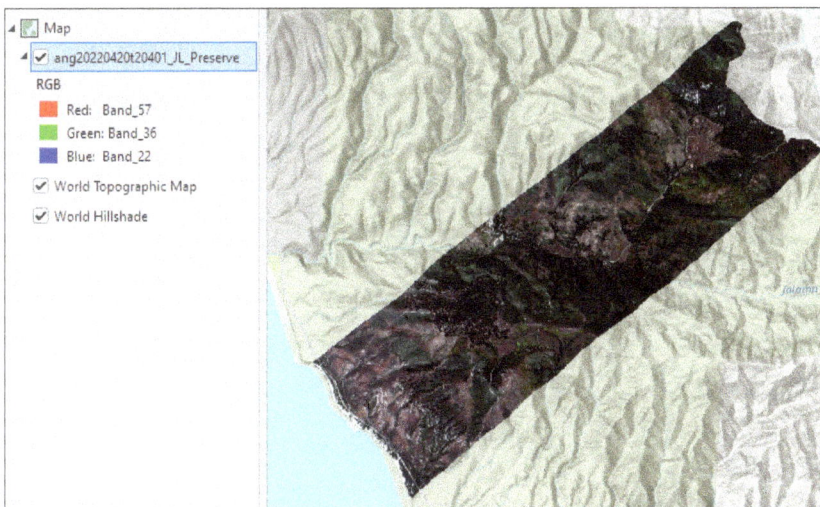

In the map, you will see a natural color display of the data with Band_57 displayed in the red channel, Band_36 displayed in the green channel, and Band_22 displayed in the blue channel.

5. In the **Contents** pane, right-click **Red: Band_57** to see the wavelength corresponding to this band.

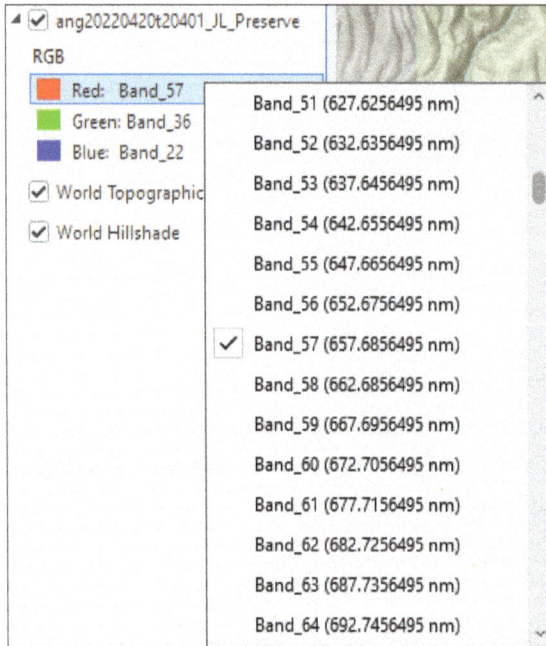

Note: The bandwidth for each band collected by AVIRIS is approximately 10 nanometers (nm). The values displayed here are the center value of the band. In other words, Band_57 spans 5 nm on either side of the listed center point and is collecting energy values in roughly the middle of the red portion of the spectrum. The red portion of the electromagnetic spectrum, by convention, spans from 600 nm to 700 nm.

Select bands using the wavelength spectrum

Because there are so many bands, you'll explore these bands in a few ways.

1. In the **Contents** pane, ensure that the **ang20220420t20401_JL_Preserve** raster layer is selected.

2. Click the **Raster Layer** tab. In the **Rendering** group, click **Band Combination** and then click **Color Infrared**.

    You will see a color infrared display with **Band_87** displayed in the red channel, **Band_57** displayed in the green channel, and **Band_36** displayed in the blue channel.

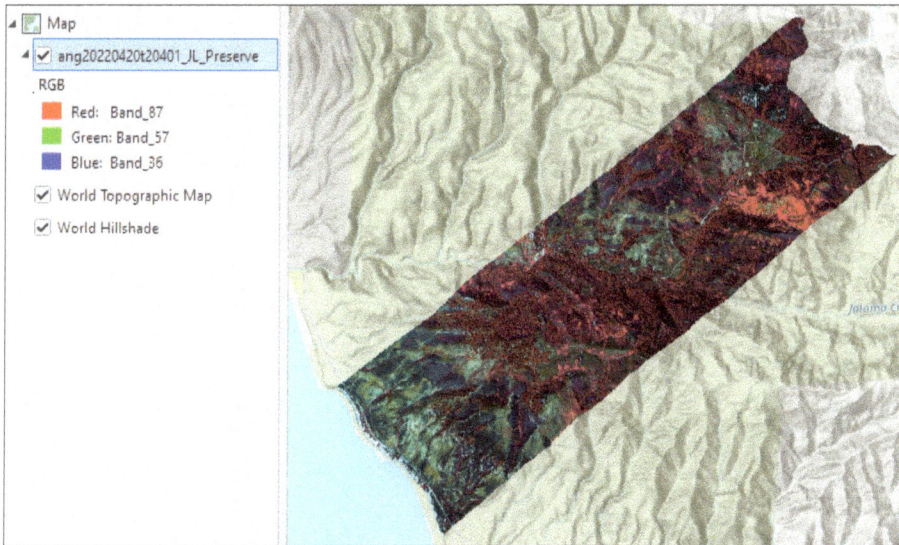

    Note: Healthy vegetation has a high reflective component in the near-infrared region of the electromagnetic spectrum, so vegetation, including trees, grass, and shrubs, all shows red hues in the image.

3. On the **Raster Layer** tab, in the **Rendering** group, click **Symbology**.

4. At the top of the pane, click the **Advanced symbology options** tab.

    You'll see the spectrum bar and three RGB handles. You can move the RGB channel handles to select bands based on wavelength.

5. Click **Select By Wavelength (nm)** and adjust the channels and bands as follows:
    *   **Red**: 1,554 nm
    *   **Green**: 1,194 nm
    *   **Blue**: 578 nm

6. Click **Apply** to update the display.

7. On the **Raster Layer** tab, in the **Rendering** group, click **DRA**.

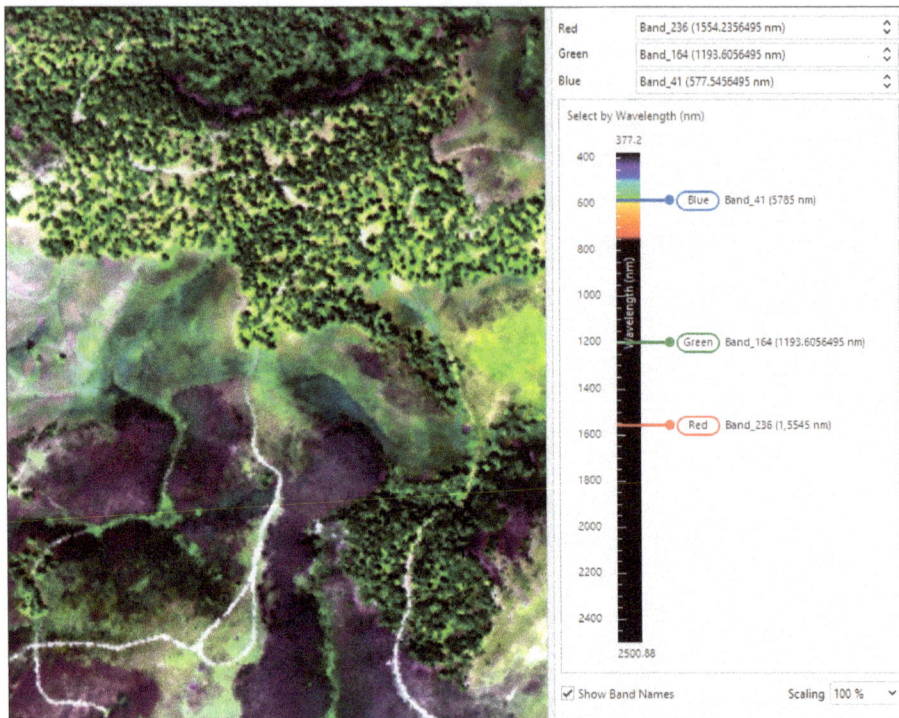

You can see that oak trees stand out as green because the oak tree leaves have higher reflectance in the Near-Infrared band (the green channel used in the display) than other types of vegetation, such as grass.

Visualize the spectral profile of the image

In this section, you'll create a spectral profile chart to visualize spectral signatures of different vegetation. You can choose to visualize the spectrum of a single pixel or the average spectra of an area.

1. In the **Contents** pane, right-click **ang20220420t20401_JL_Preserve**. Hover over **Create Chart** and click **Spectral Profile**.

    The chart window and chart pane appear.

2. In the **Chart Properties** pane, under **Define an area of interest**, click the **Point** tool.

3. On the map, click a location to place a point and draw a spectral profile.

4. Repeat this operation using point, line, or other methods to collect additional spectral signatures.

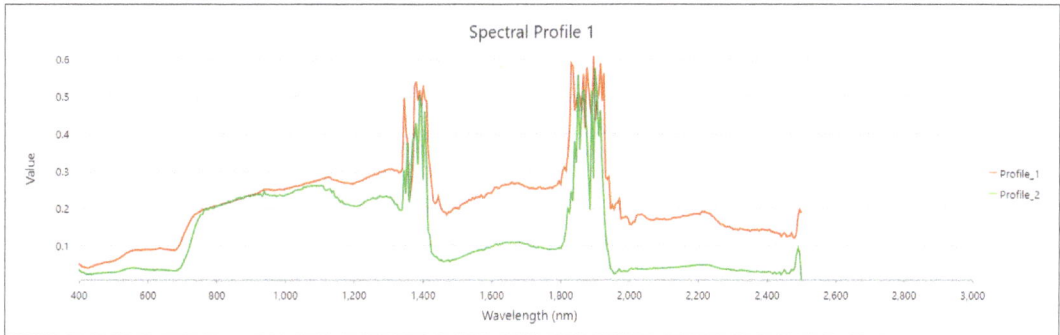

The spectral profiles of these locations are plotted in the chart. The spikes around wavelength ranges centered around 1,300 and 1,900 nm represent noisy bands caused by water vapor in the atmosphere. These bands can be excluded during your analysis.

5. Close the chart window.

Tutorial 15-2: Perform analysis

In this tutorial, you will collect some oak tree samples using the **Training Samples Manager** and then identify oaks in the image by using the **Classify Raster Using Spectra** geoprocessing tool. You will also remove the noisy bands using the **Subset Bands** function before analysis.

Collect oak tree image training samples

In this section, you will collect a few oak tree samples.

1. On the **Imagery** tab, in the **Image Classification** group, click **Classification Tools** and then click **Training Samples Manager**.

    The **Image Classification** pane appears.

2. In the **Image Classification** pane, click the **Classification Schema** list and click **Browse to existing schema**.

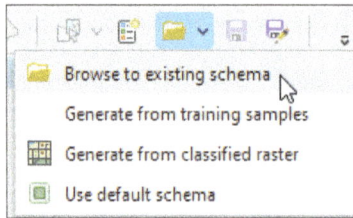

3. In the **Load Schema** dialog box, browse to **C:\Top20Imagery\Top20Imagery_ 15** and click **Hyperspectral_Oaks.ecs**. Click **OK**.

   The **Hyperspectral_Oaks** schema is loaded into the **Classification** pane, containing one class: **oaks**.

4. In the **Image Classification** pane, click the **Oaks** class and then click the **Point** tool.

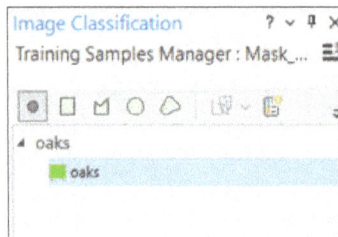

5. In the map, pan and zoom in to an area containing oak trees (dark green), and then click on oak trees to select several training samples.

6. In the lower panel of the **Image Classification** pane, click the **Save** button to save the point training samples as a feature class named oak_tree_samples. Click **Save**.

Preprocess hyperspectral imagery

In this section, you will remove the noisy bands and then classify the oak trees in the image.

1. On the **Imagery** tab, in the **Analysis** group, click **Raster Functions**.

2. At the top of the **Raster Functions** pane, click the search bar and type Subset Bands. Click the **Subset Bands** raster function result.

    You want to exclude the bands associated with atmospheric absorption wavelengths from 1338.86 to 1454.05 and from 1799.06 to 1989.9.

3. In the **Subset Bands Properties** tool, specify which bands to include and which to exclude by applying the following settings:
    *   **Raster**: ang20220420t20401_JL_Preserve
    *   **Selection Method**: Band Wavelengths
    *   **Minimum**: 377.19565
    *   **Maximum**: 382.20565
    *   **Combination**: 377-1338 1454.1-1799 1990-2500
    *   **Missing Band Action**: Best Match

4. Click **Create new layer**.

    A raster function layer is added to the map without the identified noisy bands. This rendering may take time.

5. Make sure the rendering for this layer is set to **Natural Color**.

    Remind me how: On the **Raster Layer** tab, click the **Band Combination** button.

Extract oak trees

You will use the **Spectral Angle Mapper** method to extract pixels that are spectrally close to oak tree training samples.

1. On the **Analysis** tab, in the **Geoprocessing** group, click the **Tools** button.

2. In the **Geoprocessing** pane, search for and open the **Classify Raster Using Spectra** tool. Apply the following settings:
   - **Input Raster**: Subset Bands_ang20220420t20401_JL_Preserve
   - **Spectra or Points**: oak_tree_samples
   - **Method**: Spectral Angle Mapper
   - **Thresholds**: 0.11
   - **Output Classified Raster**: ang20220420t20_oaks.crf

3. Click **Run**.

   The output classified raster is generated, added to the map, and listed in the **Contents** pane.

Note: Until the **Classify Raster Using Spectra** tool is run, an appropriate **Thresholds** parameter value is unknown. After running the tool the first time and examining the output score raster together with the source imagery and training sample data, the **Thresholds** value of 0.11 was determined. Then the tool is run a second time with the **Thresholds** value of 0.11, shown in the previous step.

Because you have only one class, **oaks**, the output score raster has only one band.

4. In the **Contents** pane, right-click the color box next to **Undefined** and click **No color**.

5. Visually verify where the classified oak trees correspond with oaks in the image. Overlay the classification result on top of the natural color image display.

    Tip: On the **Raster Layer** tab, use **Flicker** or **Swipe** to compare the two layers.

Summary

In this chapter, you learned how to work with hyperspectral imagery using ArcGIS Pro. You added hyperspectral data to a map, explored spectral bands, and created a few spectral profiles. You also learned how to analyze hyperspectral imagery and identify features of interest in the data.

Information at your fingertips

The following links are useful for providing information about concepts and tools for working with hyperspectral imagery:

- Understand hyperspectral imagery in ArcGIS: links.esri.com/Hyperspectral
- An overview of supported hyperspectral sensors in ArcGIS: links.esri.com /Supported_HS_sensors
- An overview of hyperspectral tools in ArcGIS: links.esri.com /HyperspectralTools

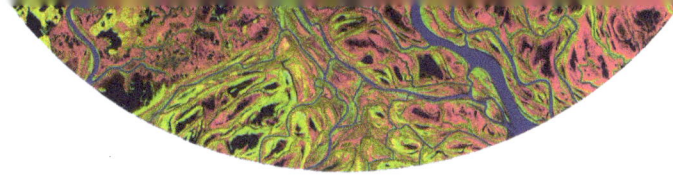

CHAPTER 16
Analyzing synthetic aperture radar

Elizabeth Ashley Menezes

Objectives

- Extract water bodies from SAR imagery for coastline mapping.
- Compute SAR indexes, such as the Radar Vegetation Index (RVI), for vegetation analysis.
- Explain the principles of SAR imagery and its applications.

Introduction

Synthetic aperture radar (SAR) uses radar signals to capture detailed images of the earth's surface, regardless of time of day or weather conditions, making it an ideal tool for flood mapping and vegetation analysis. It also provides the opportunity to study places, such as polar regions, which are dark during the winter months, or tropical regions, which have high moisture and cloud coverage that often interfere with optical satellite observations.

The satellite's SAR sensor transmits pulses of radiofrequency energy to detect objects and determine the distance (or ranging) relative to the satellite. Radio frequency is a broad type of electromagnetic energy that contains microwaves. The wavelength of the signal determines the penetration depth. For example, longer wavelengths can penetrate deeper into terrain, allowing the sensor to collect data through vegetation. Longer wavelengths can also pass through obstacles, including clouds, smoke, and rain.

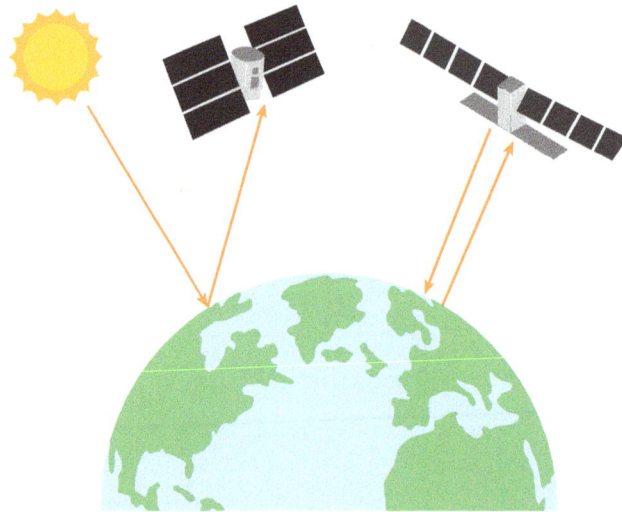

Figure 16-1. An optical sensor (*left*) can collect imagery only with clear, sunny skies. A SAR sensor (*right*) can operate effectively and collect data day and night, even when clouds are present. Courtesy of Elizabeth A. Menezes.

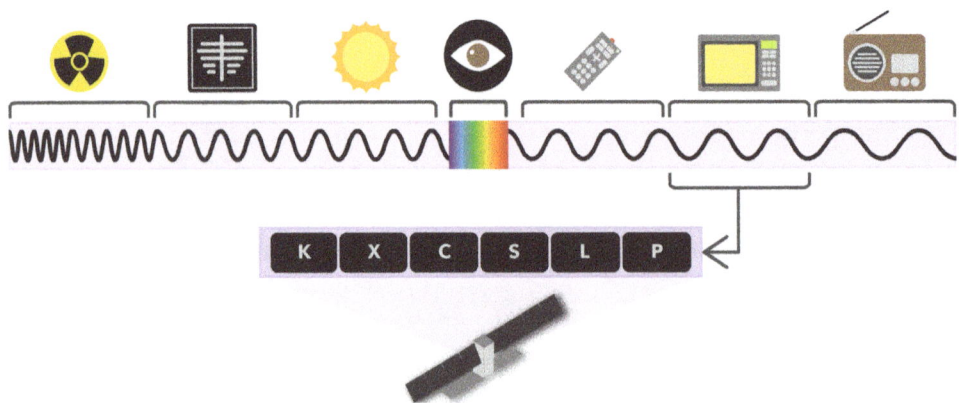

Figure 16-2. Electromagnetic spectrum showing short to long wavelengths and the respective types of waves (*left to right*: gamma rays, X-rays, ultraviolet, visible, infrared, microwave, radio), as well as the bands that SAR sensors can use within the microwave range. Courtesy of Elizabeth A. Menezes.

Note: Certain geoprocessing tools require specific radiometric terrain correction (RTC) processing. The following tutorial uses Sentinel-1 SAR Ground Range Detected (GRD) dual-polarized data from the European Space Agency (ESA) over Balikpapan Bay in Indonesia, acquired on April 1, 2018. This data is precise orbit corrected and has gone through the required RTC processing steps.

In this chapter, you'll learn how to work with SAR imagery to create informative and visually appealing maps. This tutorial will teach you how to create color composites, extract water bodies, and compute SAR indexes. You'll be equipped to apply these techniques to real-world scenarios and know how to process and interpret SAR data to create meaningful visualizations and extract valuable information. These skills are useful in the real world for tasks such as monitoring natural disasters, managing water resources, and assessing vegetation health.

Tutorial 16-1: Extract water bodies from SAR imagery for coastline mapping

In this tutorial, you will learn how to extract water bodies from SAR imagery for coastline mapping. This process is useful for identifying water bodies, which can help in monitoring coastal change and managing water resources.

By the end of this tutorial, you will be able to create detailed maps that clearly show the boundaries between land and water, enhancing your ability to analyze and manage coastal areas effectively.

Download the tutorial data and set up the project

You'll download a project that contains all the data for the following tutorials and open it in ArcGIS Pro.

1. Go to links.esri.com/Imagery20Data and download the data for chapter 16.

2. Unzip the folder to **C:\Top20Imagery**.

> **Note:** In the second chapter, you created a folder named **Top20Imagery** on your C: drive. If you haven't done that, create that folder now. Now and in subsequent chapters, you will download and unzip the data for each chapter to this folder.

3. Inside the **Top20Imagery_16** folder, double-click **Top20Imagery_16.aprx** to open the ArcGIS Pro project for this chapter.

Run the Extract Water tool

You will run the **Extract Water** tool using the prepared SAR data and a digital elevation model, or DEM, to classify pixels as water or land based on radar backscatter.

1. On the **Analysis** tab, in the **Geoprocessing** group, click **Tools**.

2. On top of the **Geoprocessing** pane, type Extract Water in the search bar.

3. Click the **Extract Water** tool.

4. In the **Extract Water** tool pane, apply the following settings:
   - **Input Radar Data**: S1_GRD_Orb_TNR_CalB0_RTFG0_Dspk.crf
   - **Output Feature Class**: S1_GRD_Orb_TNR_CalB0_RTFG0_Dspk_EW
   - **Minimum Area**: 50000 Square Meters

5. For **DEM Raster**, click the **Browse** button. In **Top20Imagery > Top20Imagery_16**, click **DEM.tif**. Click **OK**.

6. Ensure the box for **Apply geoid correction** is checked.

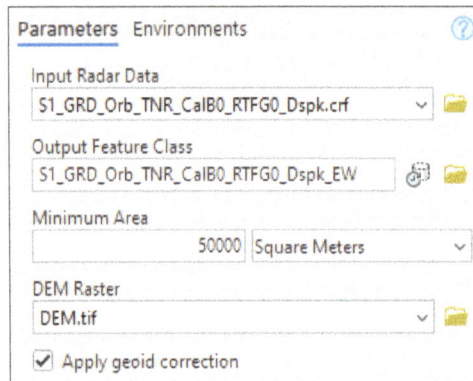

7. Click **Run**.

   This tool creates polygons for water and for areas that are not water, which will be considered land areas, making it easier to delineate coastlines.

Extract Water tool

The **Extract Water** tool is designed to help you identify and classify water bodies from SAR imagery, which is useful for coastline mapping. This process involves using prepared SAR data and a DEM to classify pixels as either water or land based on radar backscatter, enhancing your ability to analyze and manage coastal areas effectively.

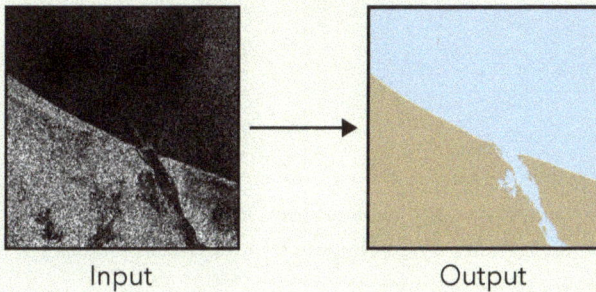

Input   Output

The tool uses the input radar backscatter from a SAR image to determine whether pixels should be classified as water (*left*) and then creates polygons for water areas. The tool also creates polygons for areas that are not water, which are considered land areas (*right*). Courtesy of Elizabeth A. Menezes.

It is important to calibrate the input radar data to gamma nought using the **Apply Radiometric Calibration** tool. This optimizes delineation and classification, especially in large radar scenes. The input radar data may not align with the output feature class. If the input radar data is not orthorectified, the tool transforms the **Output Feature Class** parameter value using the **DEM Raster** parameter. When no DEM is provided, the tool performs a transformation using an ellipsoidal surface. For optimal transformation of the **Output Feature Class** parameter value, you'll need to provide an input DEM for the **DEM Raster** parameter. The input DEM must be in the WGS84 (EPSG:4326) geographic coordinate system.

Symbolize the results

Now that you have the output, you'll symbolize the polygons according to the classification.

1.   In the **Contents** pane, click **S1_GRD_Orb_TNR_CalB0_RTFG0_Dspk_EW**.

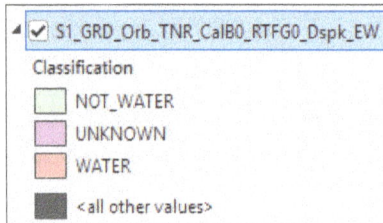

> **Tip:** If the output layer does not show three classifications, right-click the layer and click **Symbology**. Change the **Primary symbology** to **Unique Values**. Under the **Classes** tab, click the **Add all values** button.

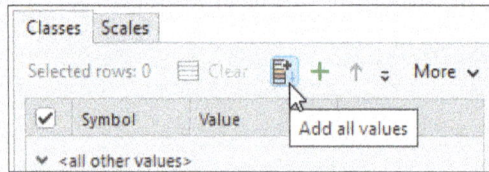

2.   On the **Feature Layer** tab, in the **Drawing** group, click **Symbology**.

3.   Click the symbol tile for the **NOT_WATER** label.

4.   On the **Properties** tab, under **Appearance**, for **Color**, expand the color palette and select **Medium Olivenite**.

5.   Using the skills you just learned, change the color for the **UNKNOWN** label to **Gray 30%** and the **WATER** label to **Moorea Blue**.

6.   On the map, you should now be able to distinctly view the symbolized results showing water as blue and land as green.

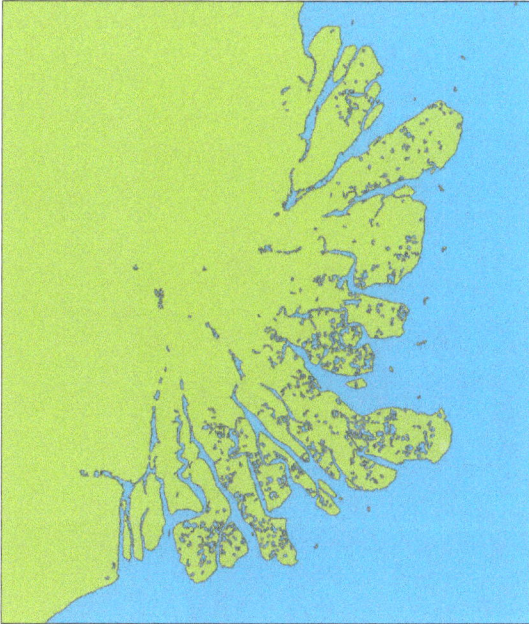

Tutorial 16-2: Compute SAR indexes for vegetation analysis

In this tutorial, you will learn how to compute the SAR indexes, using the Radar Vegetation Index, or RVI, for vegetation analysis. This process is useful for assessing vegetation growth and can help in monitoring changes in forest cover.

Throughout this tutorial, you will process SAR data and compute the RVI, which is the ratio of cross-polarized backscatter to the total backscatter from all polarizations. This index is useful for gaining insights into vegetation growth, enabling you to make informed decisions based on your analysis.

Run the Compute SAR Indices tool

You will run the **Compute SAR Indices** tool by using the prepared SAR data, which will automatically determine the appropriate index formula to use for the selected index and the input polarizations available.

1. On the **Analysis** tab, in the **Geoprocessing** group, click **Tools**.

2. At the top of the **Geoprocessing** pane, type Compute SAR Indices in the search bar.

3. Click the **Compute SAR Indices** tool. In the tool, apply the following settings:
    - **Input Radar Data**: S1_GRD_Orb_TNR_CalB0_RTFG0_Dspk_GTC.crf
    - **Output Raster**: S1_GRD_Orb_TNR_CalB0_RTFG0_Dspk_GTC_RVI.crf
    - **Index**: Radar Vegetation Index (RVI)

4. Click **Run**.

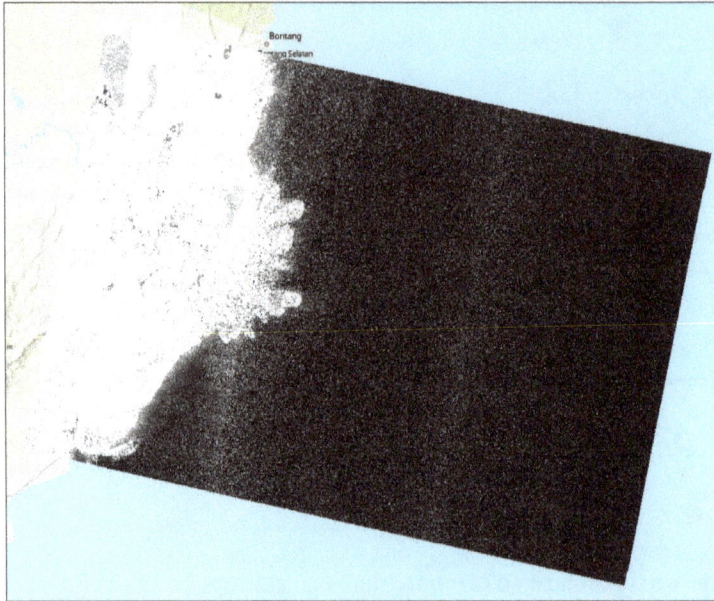

The **Radar Vegetation Index (RVI), Radar Forest Degradation Index (RFDI),** and **Canopy Structure Index (CSI)** are all available to choose from. The formulas used for these indexes depend on the polarizations offered in the radar data.

Radar Vegetation Index (RVI)

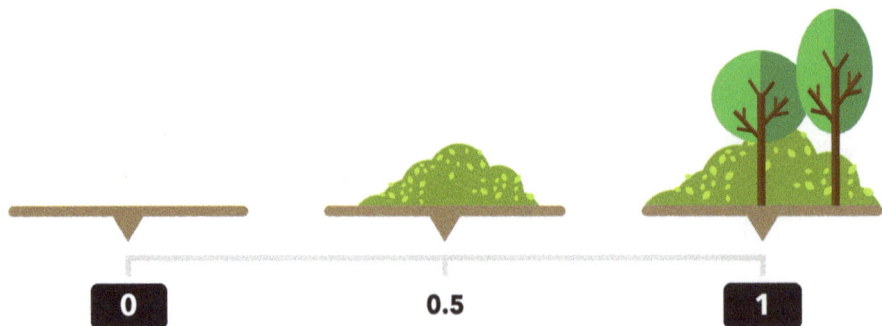

For the RVI, the values range between 0 and 1, where values near 0 indicate barren landscapes and values near 1 represent vegetated landscapes. Courtesy of Elizabeth A. Menezes.

Symbolize the results

Now that you have the output, you'll symbolize the polygons according to their classification.

1. In the **Contents** pane, you will select the **S1_GRD_Orb_TNR_CalB0_RTFG0_Dspk_GTC_RVI** layer.

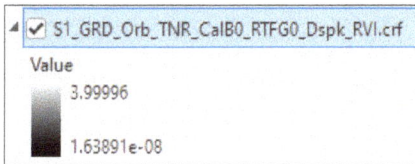

2. On the **Raster Layer** tab, in the **Rendering** group, click **Symbology**.

3. In the **Symbology** pane, apply the following settings:
    *   **Stretch type**: Minimum Maximum
    *   Check the box for **Edit min/max**.
    *   **Value**: 0 to 1
    *   **Label**: 0 to 1

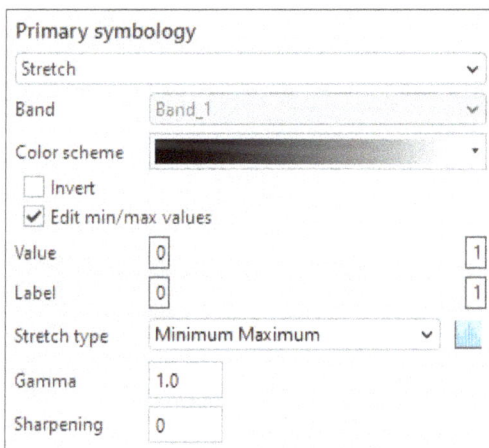

The range values and label are now 0 to 1 so that RVI values near 0 indicate barren landscapes whereas larger values indicate vegetated landscapes.

4. For **Color scheme**, expand the list and check the **Show names** box. Select the **Green Light to Dark** color scheme.

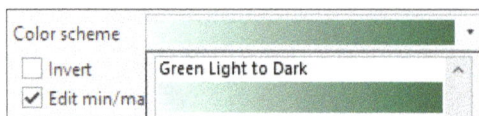

On the map, the layer updates to the new symbology where the light-green colors indicate barren landscapes and the darker green colors indicate vegetated landscapes.

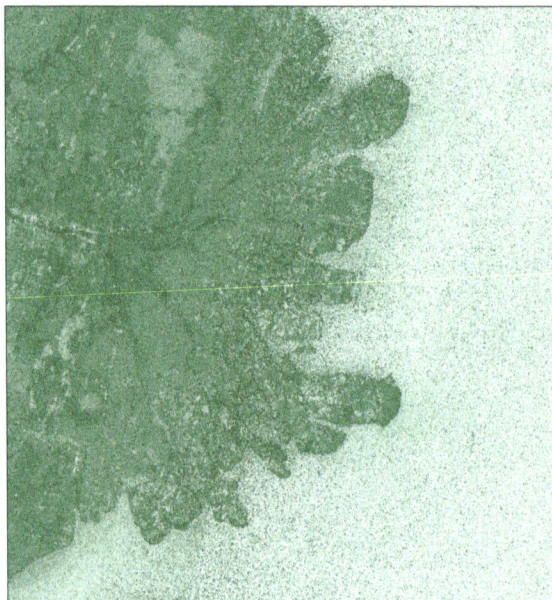

Summary

In this chapter, you learned how to work with SAR imagery to create informative and visually impactful maps. The tutorials included in this chapter teach you how to extract water bodies for coastline mapping and compute SAR indexes, such as the Radar Vegetation Index, for vegetation analysis. These skills are valuable for real-world tasks, such as managing water resources and assessing vegetation health.

Information at your fingertips

Fundamentals

Satellites with optical sensors use passive sensing, where the sunlight is reflected on the earth's surface and is sensed by a satellite. Satellites with SAR sensors use active sensing, where the sensor sends signals to the earth's surface and captures the reflected signals that travel back to the sensor, which are called the measured backscatter. The sensor uses the measured backscatter portion of the signal to render a two-dimensional SAR image.

Processing history

To find out more about the images used in this tutorial, in the Contents pane, right-click the image name and click Properties. In the Layer Properties window, click Source and expand Processing History to find information about the preparatory processing that was applied to obtain these analysis-ready SAR images.

Additional SAR indexes

The Radar Vegetation Index (RVI), Radar Forest Degradation Index (RFDI), and Canopy Structure Index (CSI) are popular SAR indexes used by the community.

The RFDI is the normalized difference between co- and cross-polarized backscatter. Lower RFDI values (less than 0.3) indicate a denser forest. Moderate RFDI values (between 0.4 and 0.6) correspond to degraded forests. Higher RFDI values (greater than 0.6) indicate deforested landscapes.

Radar Forest Degradation Index (RFDI)

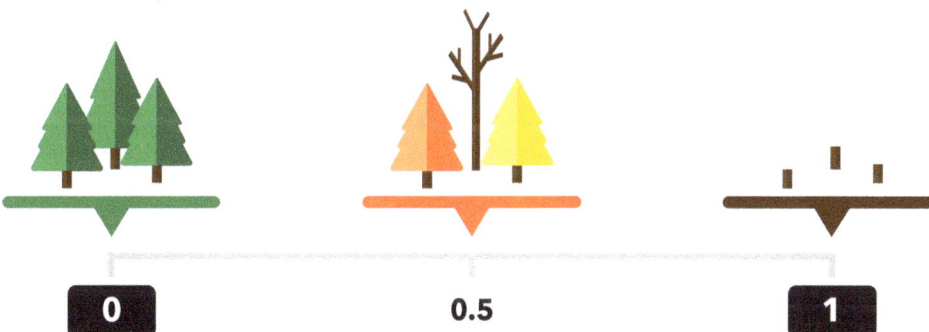

For the RFDI, the values range between 0 and 1, where values near 0 indicate a healthy forest and values approaching 1 represent degraded forests. Courtesy of Elizabeth A. Menezes.

The CSI is the normalized difference of copolarized backscatter (HH, VV), both of which are usually found in quad-polarized datasets.

Canopy Structure Index (CSI)

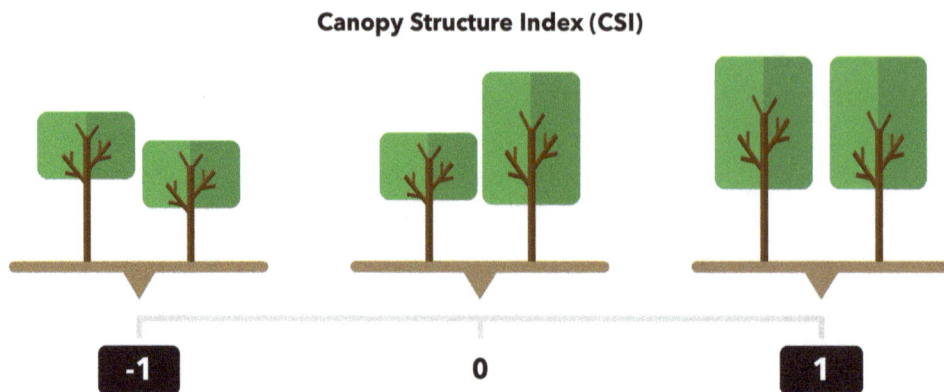

For the CSI, values range between −1 and +1, where canopies dominated with vertical structures have CSI values near −1, whereas those dominated by horizontal structures have CSI values near 1. Courtesy of Elizabeth A. Menezes.

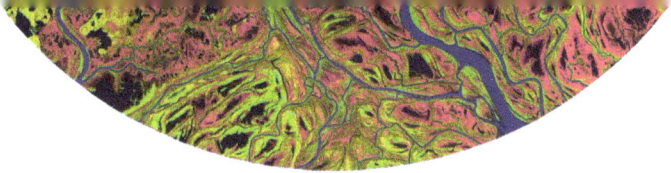

CHAPTER 17
Working with image services and online archives

Jeff Swain and Jeff Liedtke

Objectives

- Learn about image services and their types.
- Create an ArcGIS Cloud Storage connection file.
- Create a SpatioTemporal Asset Catalog connection and access these catalogs.
- Access collections from the Microsoft Planetary Computer Data Catalog and add it to a mosaic dataset in ArcGIS Pro.
- Access and use data from ArcGIS Living Atlas of the World.

Introduction

ArcGIS Image enables your organization to host, analyze, and stream imagery and raster data from the ArcGIS Pro desktop, on-premises server infrastructure, or the cloud. This chapter focuses on the integration of ArcGIS Pro with other supported services, including ArcGIS Image Server, the ArcGIS Enterprise portal, SpatioTemporal Asset Catalog (STAC), ArcGIS Microsoft Planetary Computer (AMPC), ArcGIS Living Atlas, and ArcGIS image services.

The various ArcGIS platforms, services, and supported archives are introduced in this section.

Image services

Image services are the raster and image data that are shared online as web services or layers using the ArcGIS Image Server extension and ArcGIS Enterprise portal. The source data could be from a raster dataset (imagery captured using satellites, aerial drones, and multidimensional), a mosaic dataset, or a layer file pointing to a raster. The main advantage of using image services is their flexibility, scalability, and capability to analyze the data dynamically without having to worry about data duplication and the processes involved. All you need is a computer terminal to access the data, which is particularly useful for remote locations.

In addition to accessing the imaging capabilities through an ArcGIS Image Server connection or a REST developer's environment, you can choose to publish an image service with the Open Geospatial Consortium (OGC), Web Map Service (WMS), or Web Coverage Service (WCS) capabilities and access them through various client applications, including web, desktop, and mobile. When publishing an image service, you can define the capabilities of how a client connects to the service.

Image services are mainly of two types: dynamic and cached image services. Dynamic image services are useful if you need to analyze the data by applying custom processing whereas the cached services retrieve pregenerated tiles from the cache. It is much faster to visualize the cached image services since the server does not need to generate the image dynamically.

SpatioTemporal Asset Catalog

STAC simplifies how you access and manage large volumes of geospatial data. Instead of storing terabytes of imagery on local servers, STAC enables you to efficiently search and access data archives directly from the cloud. STAC brings together a wide range of geospatial assets from various data providers into a single, standardized interface.

When accessing secure datasets stored online, you'll need an ArcGIS Cloud Storage (ACS) connection file, which securely passes your credentials to authorize access. The first tutorial shows how to create an ACS connection file. ACS connections are not required for datasets stored in publicly accessible cloud locations

ArcGIS Microsoft Planetary Computer

The AMPC is a cloud-based geospatial platform that focuses mainly on environmental sustainability and earth science. It uses open-source tools and supports open standards. It provides a vast collection of geospatial data hosted on Microsoft Azure for free to users. Most of its collections consist of various imagery types, such as Landsat, Sentinel, Moderate Resolution Imaging Spectroradiometer (MODIS), the National Agriculture Imagery Program (NAIP), Aster, and the like, which are stored in Azure Blob Storage, organized using the STAC standard. STAC provides a standardized metadata structure allowing ArcGIS Pro to connect and access the data by location, time, and other metadata.

ArcGIS Living Atlas

ArcGIS Living Atlas is an evolving digital repository assembled from the collective work of the global GIS community. It's a location where maps, applications, and data layers are available to support your work. ArcGIS Living Atlas includes data provided by users based on their research, as well as data created directly by Esri. Most data available on ArcGIS Living Atlas, however, is curated and authoritative data that is maintained to provide you with the most accurate and up-to-date information for your work. For instance, imagery from the European Space Agency (ESA) or from the US Geological Survey (USGS) and other NASA imagery besides MODIS and the Earth Observing System (EOS) are also available as image services. The data provided in ArcGIS Living Atlas is ready for you to explore to enhance your project.

ArcGIS image services

ArcGIS Living Atlas includes ready-to-use imagery and raster data layers from open-source and commercial imagery data providers, Esri partners, and worldwide sources. ArcGIS Living Atlas allows you to find the imagery content you need to visualize remotely sensed images on maps, perform image analysis, and layer imagery data with other spatial information.

Access is provided to a vast collection of imagery and remotely sensed data, including but not limited to:

- **World imagery:** Access one-meter or better aerial and satellite imagery of the world, with political boundaries and place-names.
- **Landsat:** Access the full Landsat data archive as multispectral apparent reflectance or surface reflectance imagery. This global dataset is updated daily and allows you to examine change and trends over more than 40 years.
- **Sentinel-2:** Access Sentinel-2 10- and 20-meter-resolution multispectral imagery with worldwide coverage updated every five days. See Earth's radiance and surface reflectance.

- **NAIP:** The National Agriculture Imagery Program is multispectral imagery captured during agricultural growing seasons in the continental United States. Access NAIP imagery from 2010 through 2019.
- **MODIS imagery:** Access full-resolution imagery collected by NASA's Terra (EOS AM) and Aqua (EOS PM) satellites.
- **World terrain:** Access global multiresolution elevation data services for visualization and analysis. Layers include slope, aspect, contour, and hillshade options.
- **Land cover:** Artificial intelligence models, applied to the entire Sentinel-2 scene collection for each year from 2017 to 2021, deliver a nine-class map of the earth's surface, including vegetation types, bare surface, water, cropland, and built areas.
- **HYCOM:** The HYbrid Coordinate Ocean Model, or HYCOM, is a product from the HYCOM Consortium, which forecasts global ocean conditions. Access a seven-day forecast and 30-day hindcast from HYCOM, reported in three-hour intervals.
- **Other Esri partners:** Imagery is also available from Vantor, Airbus, Planet, Nearmap, BlackSky, Impact Observatory, Vexcel Imaging, and more.

Tutorial 17-1: Use cartographic techniques to enhance a web map

MPC, STAC, and ArcGIS are tightly integrated to help you find, access, and analyze geospatial information directly in ArcGIS Pro. In this tutorial, you will fetch the Landsat Collection 2 Level-2 data, create an ACS connection file, make a STAC connection, access data from MPC through the Explore STAC pane, and add data to a mosaic dataset using ArcGIS Pro.

Create an ACS connection file

You'll create an .acs file in ArcGIS Pro for the Landsat Collection 2 Level-2 data from AMPC.

1. Open ArcGIS Pro and create a new **Map** project. Name it Top20Imagery_17 and save it to C:\Top20Imagery. Check the box to **Create a folder for this local project**.

2. On the **Insert** tab, in the **Project** group, click **Connections** > **Cloud Store** > **New Cloud Storage Connection**.

3. In the **Create Cloud Storage Connection** pane, enter the following:

- **Connection File Name**: Landsat_planetaryV2
- **Service Provider**: AZURE
- **Access Key ID (Account Name)**: landsateuwest
- **Bucket (Container) Name**: landsat-c2
- **Provider Options**:
  - **Name**: ARC_TOKEN_OPTION_NAME
  - **Value**: AZURE_SAS
  - **Name**: ARC_TOKEN_SERVICE_API
  - **Value**: https://planetarycomputer.microsoft.com/api/sas/v1/token /landsateuwest/landsat-c2
  - **Name**: ARC_DEEP_CRAWL
  - **Value**: NO

Create Cloud Storage Connection ✕

Connection File Name

Landsat_planetaryV2

Service Provider

AZURE

Authentication

Access Key ID (Account Name)

landsateuwest

Secret Access Key (Account Key)

Bucket (Container) Name

landsat-c2

Folder

Region (Environment)

Service Endpoint

Provider Options

Name		Value
ARC_TOKEN_OPTION_NAME	⌄	AZURE_SAS
ARC_TOKEN_SERVICE_API	⌄	https://planetarycomputer.microsoft...
ARC_DEEP_CRAWL	⌄	NO
	⌄	

4. Click **OK**.

This creates a cloud storage connection file, which is added to your project folder.

> **Tip:** To see how the settings in the figure were obtained, see "Information at Your Fingertips" at the end of this chapter.

Create and explore a STAC connection

1. On the **Insert** tab, click **Connections** > **STAC Connection** > **New STAC Connection**.

2. Enter the following:

   - **Connection Name**: Landsat_c2
   - **Connection**: Microsoft Planetary Computer
   - **Cloud Storage Connections (Optional)**: C:\Top20Imagery\Top20Imagery_ 17\Landsat_planetaryV2.acs

> **Tip:** To obtain the path to **Cloud Storage Connections (Optional)**, click the plus button, navigate to the project folder, select the **Landsat_planetaryV2.acs** file, and click **OK**. This file was created in the previous section.

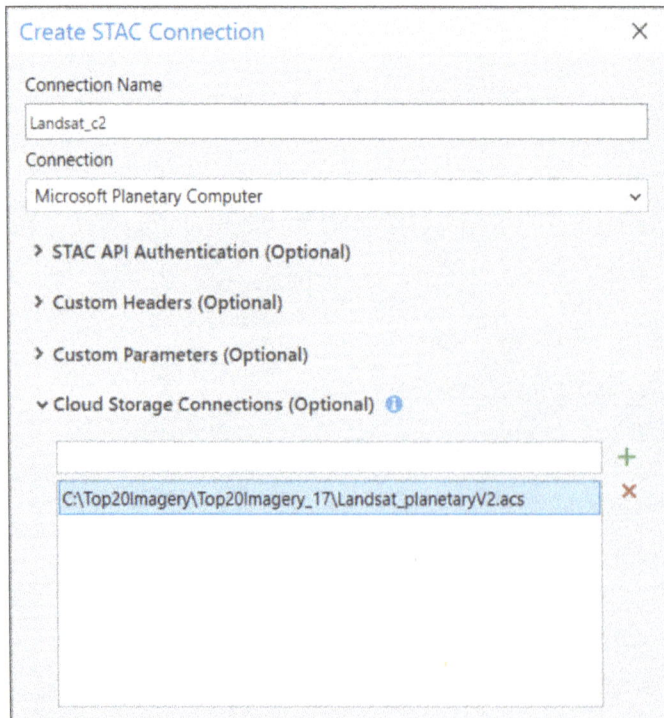

Create STAC Connection   ✕

Connection Name

| Landsat_c2 |

Connection

| Microsoft Planetary Computer   ⌄ |

> STAC API Authentication (Optional)

> Custom Headers (Optional)

> Custom Parameters (Optional)

⌄ Cloud Storage Connections (Optional) ⓘ

| | +

| C:\Top20Imagery\Top20Imagery_17\Landsat_planetaryV2.acs | ✕

3. Click **OK**.

    This creates a STAC connection (.astac file) within the project folder. You can access it from the **Catalog** pane.

4. In the **Catalog** pane, expand **STACs**, right-click the STAC connection listed, and click **Explore STAC**.

5. In the **Explore STAC** pane, on the **Parameters** tab, in the **Search Collections** field, search for landsat-c2-l2. In the results, find the **landsat-c2-l2** collection and check the box.

6. Find your own imagery of interest by adjusting the filter for **Date and Time** and select a date range of just a few days.

7. Adjust the filter for **Extent**, zoom the map to an area of interest, and click the filter for **Current Display Extent**.

8. Click **View Results**.

    The results appear on the **Results** tab.

9. Return to the **Parameters** tab and continue adjusting the filters until you have found a collection of a few images in a localized area.

    > **Tip:** On the **Results** tab, you can see the date the imagery was taken. Use this information to limit your date range filter. On each image, click **Show Footprint** to see the image footprint.

Create a mosaic dataset

You will create a mosaic dataset using imagery available on the AMPC using your STAC connection. You will create a mosaic dataset in your geodatabase and then add images to it.

1. In the **Catalog** pane, expand **Databases**, right-click **Top20Imagery_17.gdb**, and click **New > Mosaic Dataset**.

2. In the **Create Mosaic Dataset** tool, complete the following:

    *   **Mosaic Dataset Name**: Landsat_Level2
    *   **Product Definition**: Landsat 9

3. Click **Run**.

4. Return to the **Explore STAC** pane. Activate the **Results** tab, if needed.

5. From your exploration of the available imagery, check the boxes for two scenes to be added to the mosaic dataset.

6. Near the top of the pane, click the **Add to Current Map** (plus folder) list and click **Add to Mosaic Dataset**.

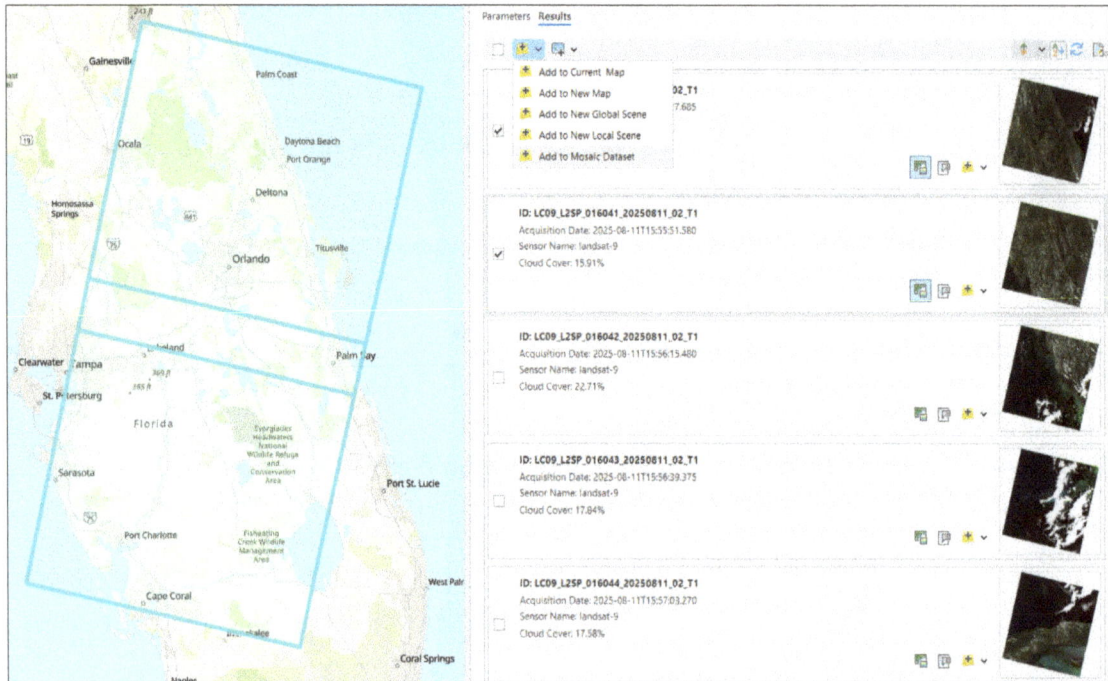

The **Add Rasters to Mosaic Dataset** tool appears with the selected scenes as **Input Data**.

7. In the **Add Rasters to Mosaic Dataset** geoprocessing tool, apply the following settings:

- **Mosaic Dataset**: Landsat_Level2
- **Processing Template**: Surface Reflectance
- Expand **Raster Processing** and check the box for **Calculate Statistics**.

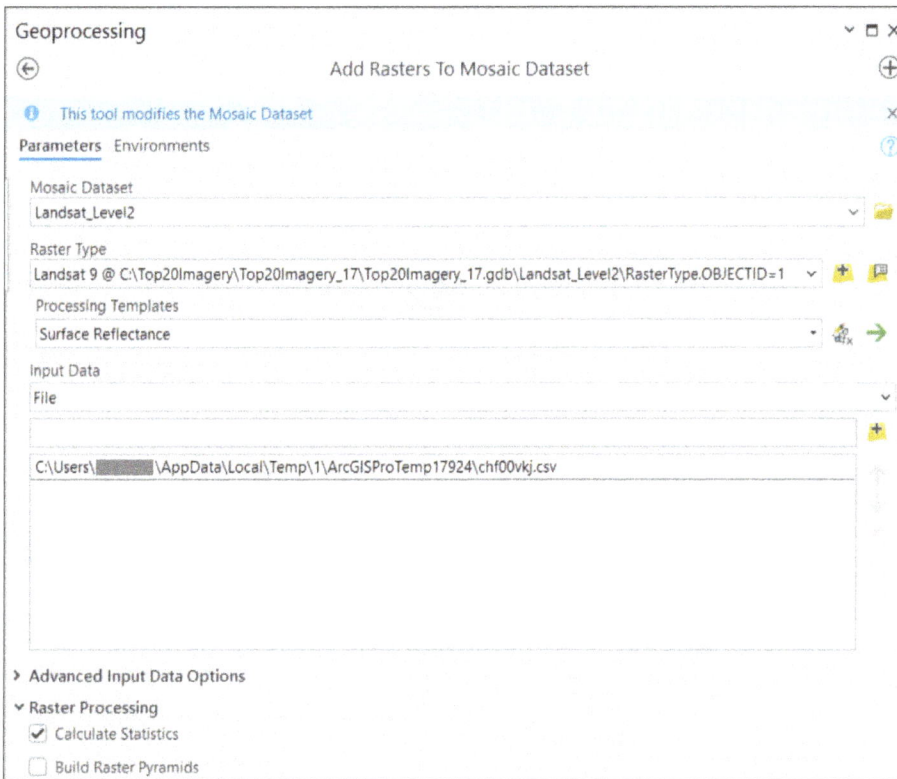

8. Click **Run**.

The selected STAC collections from AMPC are added to a mosaic dataset. The mosaic dataset is created, displayed in the map, and listed in the **Contents** pane. It is now stored locally and available for any additional workflows.

Tutorial 17-2: Access and use data from ArcGIS Living Atlas of the World

Accessing data from ArcGIS Living Atlas is possible in several ways from ArcGIS Pro. ArcGIS Living Atlas is integrated into ArcGIS Pro and appears by default in the **Catalog** pane, **Catalog** view, or from the **Add Data** dialog box. You can then use the ArcGIS Pro filtering options to find the data you are looking for.

Browse through ArcGIS Living Atlas imagery layers from the Catalog pane

ArcGIS Living Atlas provides many imagery layers for you to use in ArcGIS Pro to explore and review. You can use the Catalog pane to access ArcGIS Living Atlas imagery layers quickly and easily.

1. In your **Top20Imagery_17** project, add a new map by clicking the **New Map** button in the **Project** group on the **Insert** tab.

2. At the top of the **Catalog** pane, click **Portal** to open the portal connections that are available.

    Note: A portal connection is available when you sign in to ArcGIS Online.

    Six options across the top of the **Catalog** pane indicate which section of the portal you are accessing. ArcGIS Living Atlas can be accessed from the button farthest to the right.

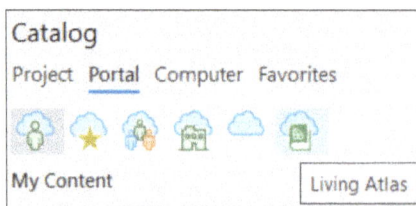

3. Click the **ArcGIS Living Atlas** button to begin browsing for imagery content.

    In the **Catalog** pane, you will see layers listed below a search bar, with a sort and filter button. The list of layers has an icon next to each layer indicating what type of layer it is. To identify the imagery layers, you will filter the list by type.

4. Click the **Filter** button, expand **Item Type > Layers**, and then click **Imagery Layers**.

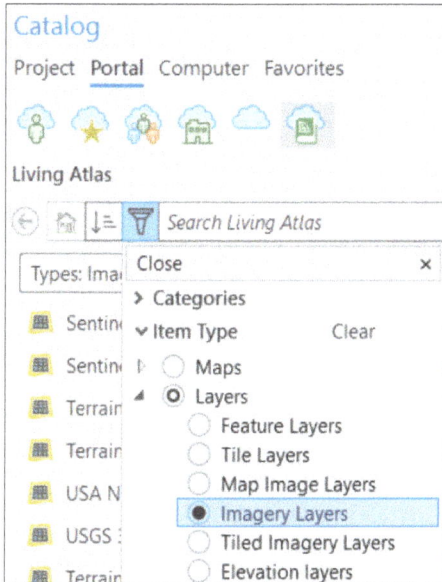

The list now displays only imagery layers.

5. Click anywhere in the **Catalog** pane to close the filter setting.

Add an imagery layer from ArcGIS Living Atlas to a map

1. Right-click the **Sentinel-2 Views** layer and click **Add To Current Map**.

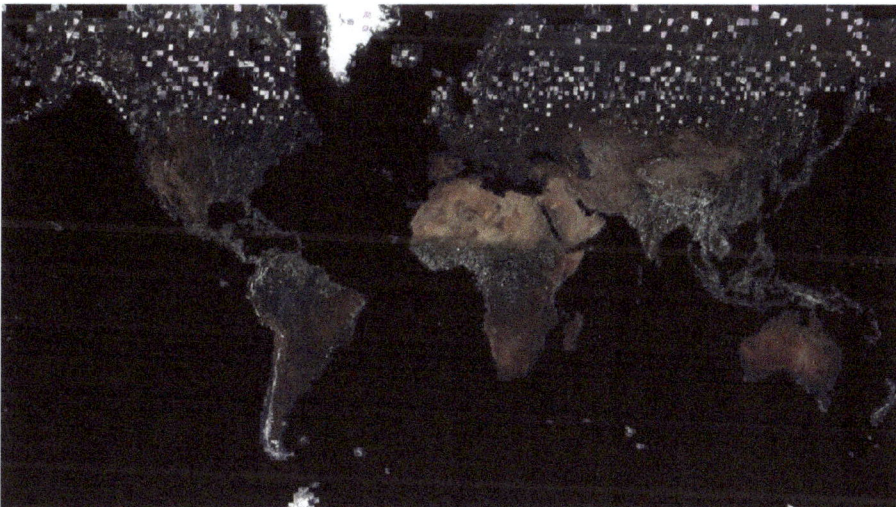

The **Sentinel-2 Views** layer is added to the map. This imagery layer contains images for multiple years and can be used to show the current view of a particular area or to search for images that meet specific criteria. Many of the imagery layers within ArcGIS Living Atlas contain much more information than what is visible initially.

2. In the **Contents** pane, right-click the **Sentinel-2 Views** layer and click **Properties** to view the layer properties.

3. In the **Layer Properties** window, click the **Source** tab and expand **Raster Information** to see the metadata of the imagery layer.

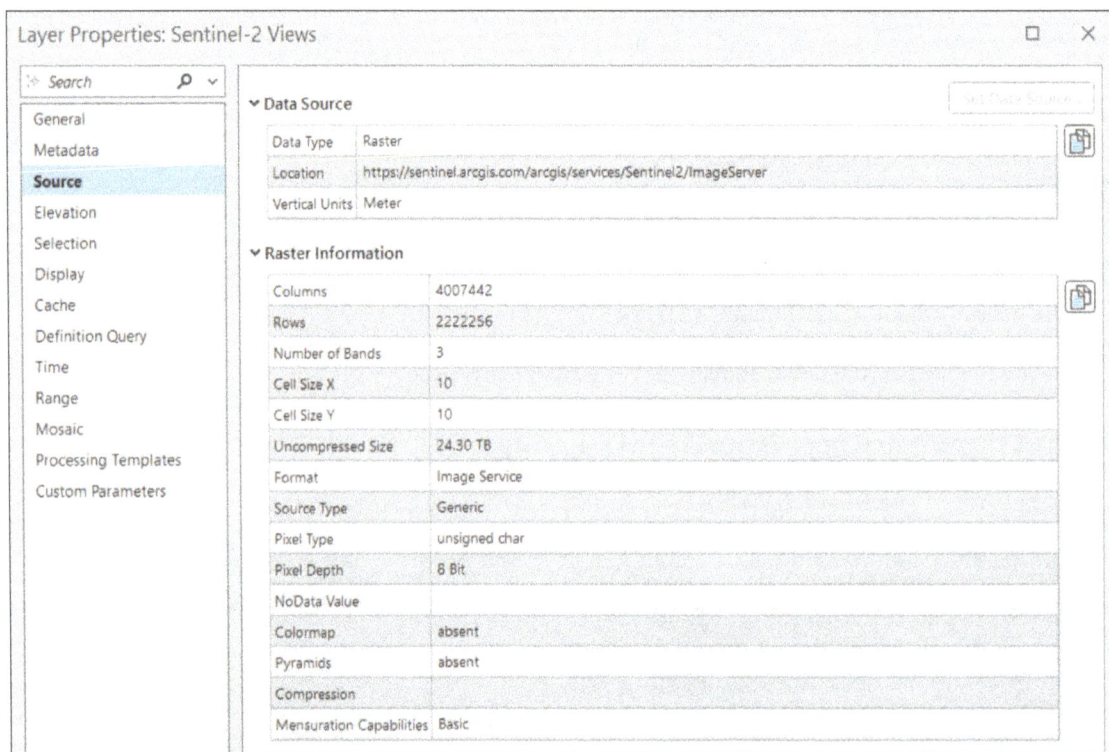

You can see the properties of the imagery layer and information about what is displayed in the map. The cell size, number of bands, and bit depth can tell you a lot about whether this imagery layer will meet your analysis or display needs.

4. Click the **Processing Template** tab and expand the list of available processing templates for this imagery layer.

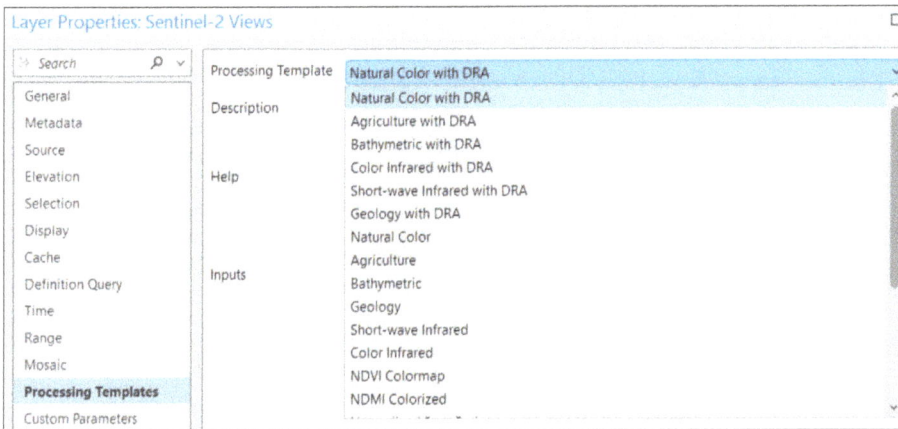

Processing templates are preconfigured raster function templates that can be used to display the imagery layer differently and aid in visualizing different phenomena. Not all imagery layers have additional processing templates, but this imagery layer contains many bands and can be used for various visualizations for your analysis.

Explore an imagery layer from ArcGIS Living Atlas in ArcGIS Pro

You will choose a visualization that uses the shortwave raster bands.

1. In the **Processing Template** menu, click **Short-Wave Infrared with DRA**. Click **OK**.

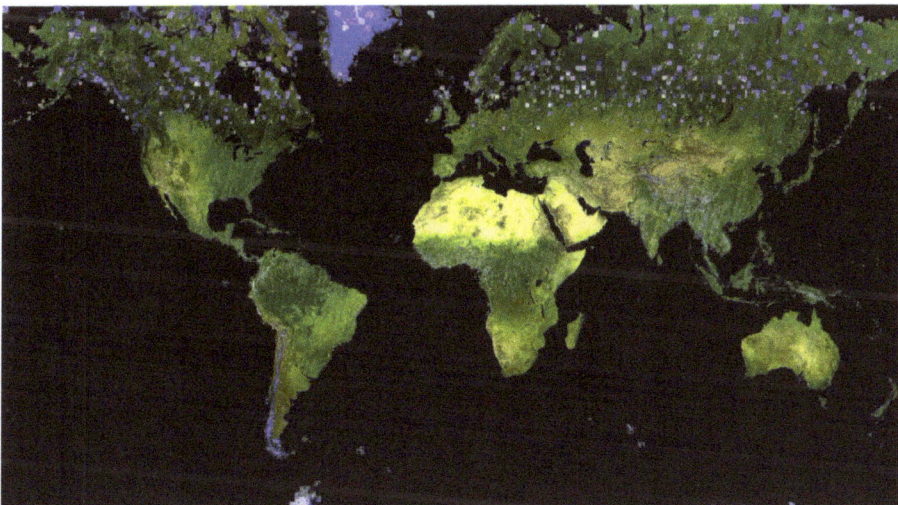

2. On the **Map** tab, in the **Inquiry** group, click **Locate** to open the **Locate** pane.

3. In the **Search** field, type Thatcham, Berkshire, England and press **Enter**.

The map zooms to the area, and you can see the imagery displayed with the new processing template. In this location, you can see the city and some of the surrounding area, as well as some water features shown as dark areas.

> **Note:** Your view of the location will vary as the imagery updates with time.

Processing templates apply the band combination and symbology choices made for that processing template, but they can also be changed.

> **Tip:** When selected, the Sentinel-2 Views layer activates the contextual **Data** tab.

4. On the **Data** tab in the **Processing** group, expand **Processing Templates**.

You can see many processing templates available for this imagery layer.

5. Click **Agriculture with DRA** to change the visualization again.

You can see the band combination change again, and now the water features, developed areas, and vegetation are all visible for further interpretation.

Imagery layers from ArcGIS Living Atlas can be visualized like other imagery layers from ArcGIS Online and, depending on the imagery layer, can also contain images from other times.

6. On the **Data** tab, click **Explore Raster Items** to open the **Raster Item Explorer** pane.

The **Raster Item Explorer** pane allows you to review information about the individual images used to create the imagery layer and will work on any dynamic imagery layer from Image Server or created as an image collection in ArcGIS Online.

Because the map is still zoomed to Thatcham, Berkshire, England, the available images are limited to this area.

7. For an image in the list, check the box and click **Properties** to see the properties of the input image used in the imagery layer.

The **Inspect** tab shows information about the input image and the metadata. Here you can also see a thumbnail of the entire image and review other properties, including the number of bands.

In addition to reviewing properties of the image this way, you can also add the input image to a map. You can customize that image's visualization or save the imagery layer for later use. This can be particularly useful if you want to maintain access to a particular image from a specific date. ArcGIS Living Atlas layers are continually being updated with the latest imagery, so if you want to maintain access to a particular image, this is a great method.

8. Click **Add to Current Map** to add this specific image to the map.

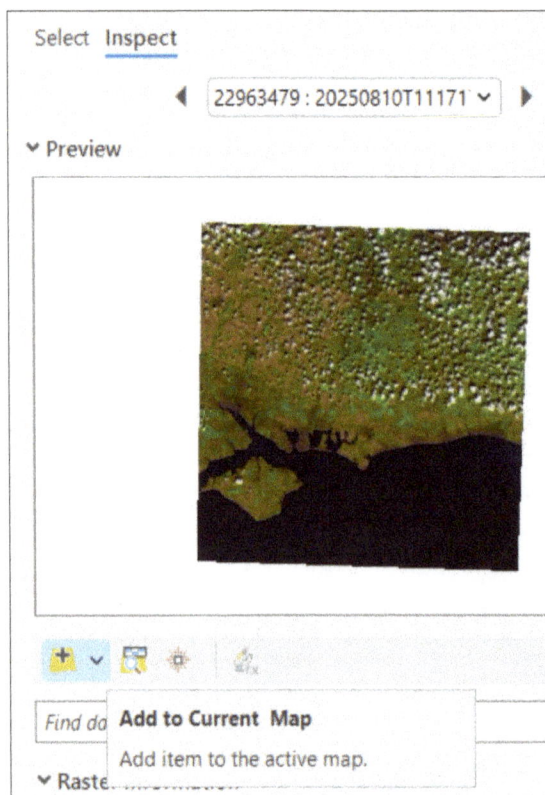

The layer appears in the **Contents** pane with the name of the individual image. The image is also visible in the map.

9. In the **Contents** pane, turn off the visibility of the **Sentinel-2 Views** layer.

10. In the **Contents** pane, right-click the new layer and click **Zoom To Layer**.

    The new layer contains all the functionality of the source imagery layer, but only the selected layer is visible.

    > **Tip:** You can save this as an item to your ArcGIS Online organization to use in other maps.

11. Using the skills you've learned, on the **Processing Templates** menu, click the **Natural Color With DRA** template and change the visualization back to natural color.

    With imagery layers from ArcGIS Living Atlas, you do not have to download the individual images to gain the functionality you are used to with local image files. You can alter band combinations and see pixel values of the imagery layers, which will allow you to use them in other raster analysis or customize the visualization for your work.

    Dynamic imagery layers can also be shared within your organization in ArcGIS Online. Those layers provide additional control over the display of the individual rasters contained within the dynamic image collection, but just as with tiled imagery layers, they can also be used for raster analysis and be visualized with different band combinations and raster function templates.

Take the next step

To get an ACS connection file, follow these steps to get the Account Name, Bucket Name, and ARC_TOKEN_SERVICE_API.

1. Select a dataset from https://planetarycomputer.microsoft.com/catalog. For example: Landsat Collection/Landsat Collection 2 Level-2.

2. Click the STAC collection link. For example: https://planetarycomputer. microsoft.com/api/stac/v1/collections/landsat-c2-l2. View the contents in your browser or in a JSON viewer.

3. Identify the Account Name and Bucket (Container) Name by searching for the metadata **msft:storage_account** and **msft:container**. The storage_account is the Account Name, and the container is the Bucket Name.

4.   Create the ARC_TOKEN_SERVICE_API URL using this template, inserting the relevant information: https://planetarycomputer.microsoft.com/api/sas/v1 /token/<storage_account >/<container>.

The URL for Landsat Collection 2 Level-2 is https://planetarycomputer.microsoft. com/api/sas/v1/token/landsateuwest/landsat-c2.

Summary

In this chapter, you created an ArcGIS Cloud Storage connection file. Next, you created a STAC connection to access a STAC catalog, accessed a collection from the MPC Data Catalog, selected imagery you were interested in, created a mosaic dataset, and accessed that mosaic dataset in ArcGIS Pro. You also learned about accessing image services in ArcGIS Living Atlas.

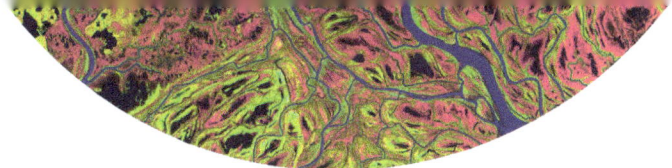

CHAPTER 18
Sharing imagery from ArcGIS Pro

Jeff Swain and Thomas Humber

Objectives

- Publish a mosaic dataset as an imagery layer.
- Adjust the rendering of an image service layer in ArcGIS Pro.
- View a newly shared imagery layer in Map Viewer.

Introduction

In this chapter, you'll learn how to publish a mosaic dataset to ArcGIS Online to create a tiled imagery layer based on the mosaic dataset you created in a previous chapter.

First, you'll explore the mosaic dataset of Central Park in New York City that comes with this chapter. This mosaic dataset includes four-band imagery of the area that you will use to review the vegetation.

Tutorial 18-1: Review and share the mosaic dataset

You'll review the imagery that was used and consider options for publishing the imagery as an imagery layer. You'll look at the source imagery properties, as well as the modifications made previously within the mosaic dataset to create the desired visualizations and renderings. Then you'll use the **Create Hosted Imagery Layer** wizard to share an imagery layer to ArcGIS Online.

> **Note:** At the end of this tutorial, you will upload a mosaic dataset. Depending on your internet connection, this may take time.

Download the tutorial data and set up the project

1. Go to links.esri.com/Imagery20Data and download the data for chapter 18.

2. Unzip the folder to **C:\Top20Imagery**.

> **Note:** In the second chapter, you created a folder named **Top20Imagery** on your C: drive. If you haven't done that, create that folder now. Now and in subsequent chapters, you will download and unzip data for each chapter to this folder.

3. Inside the **Top20Imagery_18** folder, double-click **Top20Imagery_18.aprx** to open the ArcGIS Pro project for this chapter.

Open a map in Map Viewer and explore the mosaic dataset

You will familiarize yourself with a mosaic dataset centered on Central Park in New York City, New York. The four images used to create this mosaic dataset are from the National Agriculture Imagery Program (NAIP) administered by the US Department of Agriculture (USDA). The four-band image has three visible bands and one near-infrared band. The standard band order for NAIP imagery is Red (Band 1), Green (Band 2), Blue (Band 3), and Near-Infrared, or NIR (Band 4). It's important to recognize band order for rendering appropriately or for subsequent analysis.

1. In the **Contents** pane, if necessary, expand the mosaic dataset, **Central_Park**.

In addition to the imagery, you will see the green outline of the footprint feature layer from the mosaic dataset. The footprint feature layer indicates the boundary of the individual images within the mosaic dataset.

2. Right-click the **Footprint** layer and click **Attribute Table**.

The attribute table of a mosaic dataset contains references to the source imagery and any overviews created and provides information in the fields about the zoom level, or scale, in the map each one will display. In this mosaic dataset, there are input images (the four **CentralParkQuarter** images) and overviews (the three **Ov_i01** rasters) to optimize the display.

3. Close the table.

Sometimes, mosaic datasets can also have raster function templates available to speed up rendering. These templates are stored with the mosaic dataset as processing templates. This mosaic dataset has two processing templates to aid in rendering.

4. In the **Contents** pane, click the **Central Park** mosaic dataset.

5. On the ribbon, click the **Data** tab. In the **Processing** group, click **Processing Templates** and then click **NDVI Colorized (Central Park-NYC)**.

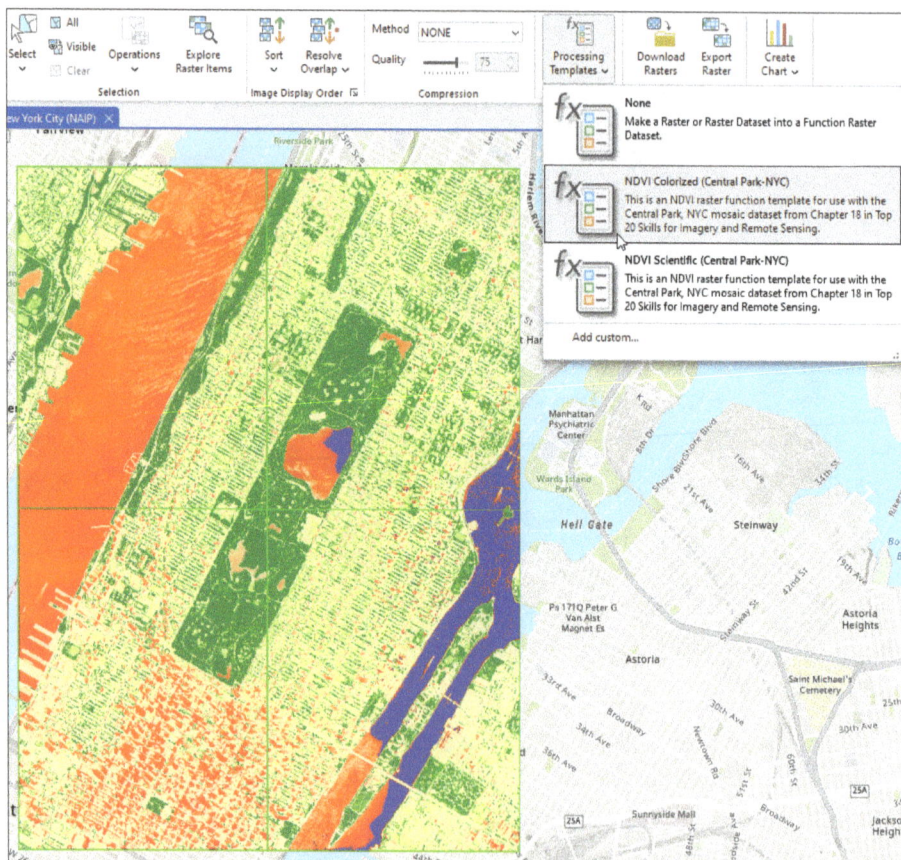

A second NDVI processing template is also available. This one renders a scientific output in grayscale values.

6. In the **Processing** group, click the **Processing Templates** and then click **NDVI Scientific (Central Park-NYC)**.

Initially, this processing template renders the image in gray scale. You need to adjust the rendering to see all the values.

7. On the **Mosaic Layer** tab, in the **Rendering** group, click **DRA**.

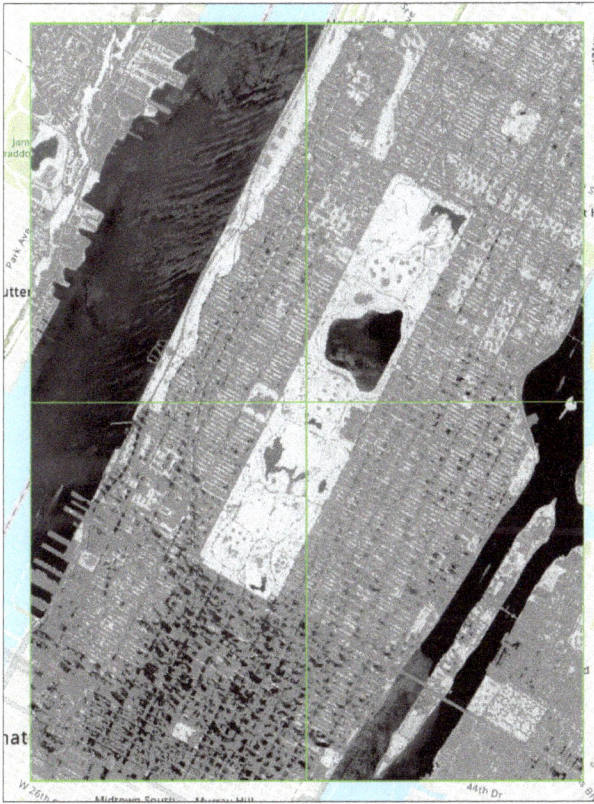

Now that you have explored the mosaic dataset and viewed some processing templates associated with it, you are ready to share it as an imagery layer in ArcGIS Online.

Open the Create Hosted Imagery Layer wizard

You can share images and mosaic datasets to ArcGIS Online directly from ArcGIS Pro. The **Create Hosted Imagery Layer** wizard can guide you through the process of sharing an imagery layer to ArcGIS Online.

> **Important:** *You must be signed in with an ArcGIS Online account that has the necessary permissions to publish imagery layers. In the ArcGIS Online help, refer to the "Publish hosted imagery layers" page.*

1. On the **Imagery** tab, in the **Share** group, click **Create Hosted Imagery** to start the process of sharing the imagery layer to ArcGIS Online.

   The first stage of the wizard offers the choice of which type of imagery layer you want to create: tiled or dynamic. Each type of imagery layer offers different capabilities for use. Depending on your workflow, you can create an imagery layer that meets your needs.

For this tutorial, you will choose a tiled imagery layer.

2. For **Choose layer type(s)**, check the box for **Tiled Imagery Layer**.

The next option defines input images you are using. The wizard allows you to choose to share a single image, a group of images, or a mosaic dataset. The options to share a single image or a collection of images are also available when sharing from a browser using ArcGIS Online. However, you can only share a mosaic dataset from the **Create Hosted Imagery Layer** wizard in ArcGIS Pro. This option allows you to customize the way your input images overlap and control other aspects of the output imagery layer by modifying the mosaic dataset properties and default settings.

3. For the question **Do you have a single image or a collection of images?**, click **Mosaic Dataset**.

Next, you'll select the input mosaic dataset.

4. Click **Next**.

Now you will define the source imagery. Because you chose the option to use a mosaic dataset, it will filter all selections based on a mosaic dataset inside a geodatabase. If you chose the other options, it would allow you to select individual images.

5. For **Source Data**, click the **Browse** button and browse to **C:\Top20Imagery \Top20Imagery_18\Top20Imagery_18.gdb**. Click the **Central_Park** mosaic dataset and click **OK**.

6. Click **Next**.

Now you will type the name of the imagery layer, add metadata, and specify a location.

7. For **Name**, type Central Park Fourband_<your initials>.

8. Add an appropriate **Description** and **Tags** for the mosaic dataset.

    If you already have folders established in your ArcGIS Online Content, you can
    save the new hosted imagery layer to one of your choice. You can also store it in
    the default location and relocate it later.

9. Click **Next,** verify the information, and then click **Run** to start the process.

    Note: Because of the size of the imagery in this mosaic dataset, the upload may take
    several minutes.

    Once the process finishes, you will see the new tiled imagery layer added to the
    map. This imagery layer can function just like a local image file, with access to
    the pixel values and raster bands to customize the visualization and use in raster
    analysis. You can also apply raster function templates—as processing templates—
    to image services in ArcGIS Pro.

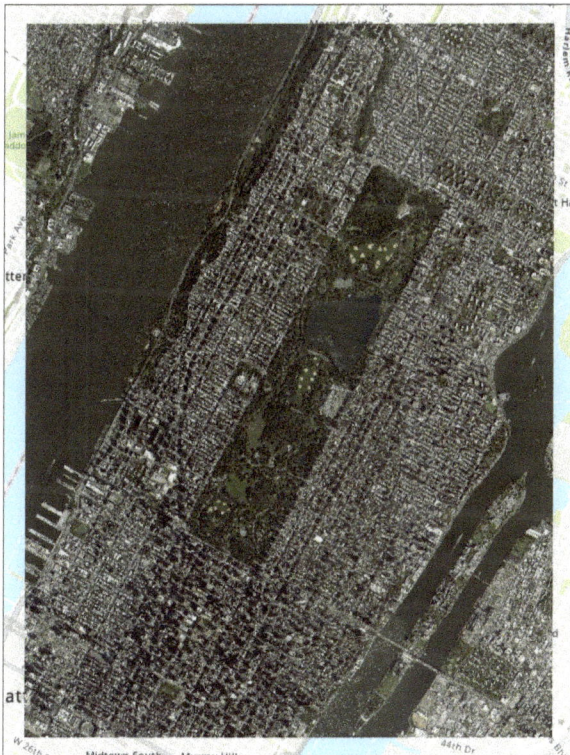

10. Click **Finish** to close the wizard.

Tutorial 18-2: Change the rendering of an image service in ArcGIS Pro

Imagery layers in ArcGIS Online function the same way as local image files in ArcGIS Pro. You can visualize them in ArcGIS Pro by changing the symbology as you would a local file. The raster band order can be manipulated, and you can also apply raster function templates to the online imagery layers.

Change the band combination

1. In the **Contents** pane, right-click the **Central_Park** mosaic dataset and click **Remove**.

2. Select the **Central_Park_Fourband** tiled imagery layer.

   Like the **Mosaic Layer** contextual tab you learned about earlier, the **Image Service Layer** contextual tab contains options available to manipulate, render, and enhance the image service layer.

3. On the **Image Service Layer** tab, in the **Rendering** group, click **Symbology** to open the **Symbology** pane.

   In the **Symbology** pane, you can see all the image bands and other options just as you would see from the locally stored images. Changing the band combination using this method relies on knowing, or remembering, the band combination order. Recall the standard band order for NAIP imagery is R,G,B,NIR. You want to render this image as a color infrared image. To do this, you'll need to set the band order to 4,1,2 (RGB).

4. In the **Primary Symbology** section, change the band combination to the following:

   - **Red:** Band_4
   - **Green:** Band_1
   - **Blue:** Band_2

Primary symbology		
RGB		⌄
Red	Band_4	⌄
Green	Band_1	⌄
Blue	Band_2	⌄
Alpha	None	⌄

Once the band order has been changed, you will see the updated visualization in the map.

The new band combination renders a color infrared visualization.

> **Tip:** You can also modify visualizations by clicking the **Band Combination** button in the **Rendering** group and selecting one of the available options. Because this image has only four bands, you have only two standardized rendering options available: **Natural Color** (1,2,3) and **Color Infrared** (4,1,2).

5. On the map, zoom in to the northern end of the park area and see the new visualization.

The color infrared band combination shows the vegetation in stark contrast to the buildings, water, and other human-made objects in and around the park. The athletic fields and softball and baseball fields are clearly visible along with other sites, such as tennis courts, basketball courts, and walking paths within the park.

6. Click a location within the imagery layer to see the pixel values.

Service Pixel Value	47, 53, 51, 128
RGB.Red	128
RGB.Green	47
RGB.Blue	53
RGB.Alpha	

Note: Your values will differ, depending on where you clicked.

The real pixel values are available for tiled imagery layers so you can use the image as part of various raster or image analysis workflows.

You can also apply any available raster function templates as processing templates in the same way you viewed them as part of the mosaic dataset.

7. On the **Data** contextual tab, in the **Processing** group, click **Processing Templates** and then click **Add Custom**.

8. In the **Select Raster Function Template** window, browse to **C:\Top20Imagery \Top20Imagery_18\RasterFunctionTemplates\Top20Imagery_18**, click **NDVI Colorized (Central Park-NYC).rft.xml**, and click **OK**.

The colorized NDVI processing template is applied to this image. You can set the rendering back by applying **None** as the processing template.

Tutorial 18-3: View new imagery in Map Viewer in ArcGIS Online

Because you shared the imagery layer to ArcGIS Online, it can be used in other applications, such as Map Viewer (in ArcGIS Online). In Map Viewer, you have many of the same visualization and raster analysis options available when using the image in ArcGIS Pro in Map view.

Use style options to change the band combination

1. Open a web browser, navigate to arcgis.com, and sign in with the credentials you used to publish the imagery layer.

2. Click the **Content** tab to see all the content you have created.

3. On the **Central Park Fourband** imagery layer item, click the **More Options** button (three dots) and click **Open In Map Viewer**.

4. Make sure the **Central Park Fourband** imagery layer is selected in the **Layers** pane.

5. On the right, on the **Settings** toolbar, click the **Styles** tab.

6. Under **RGB**, click **Style options**.

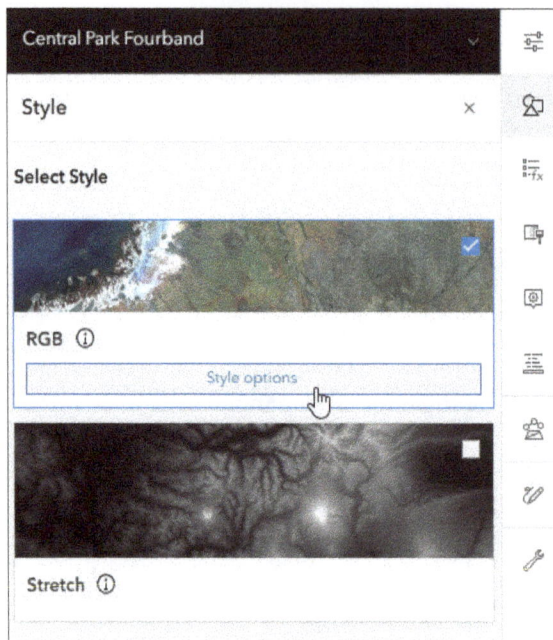

There are options to change the band combination, the stretch type, and even create your own preset style. You can style the imagery layer in many ways, depending on your interpretation needs.

7. Change the band combination to the 4,1,2 color infrared band combination you used previously. Change the **Stretch type** to **Percent clip**.

The imagery layer will update as you change the options.

8. When the style options are modified, click **Done**.

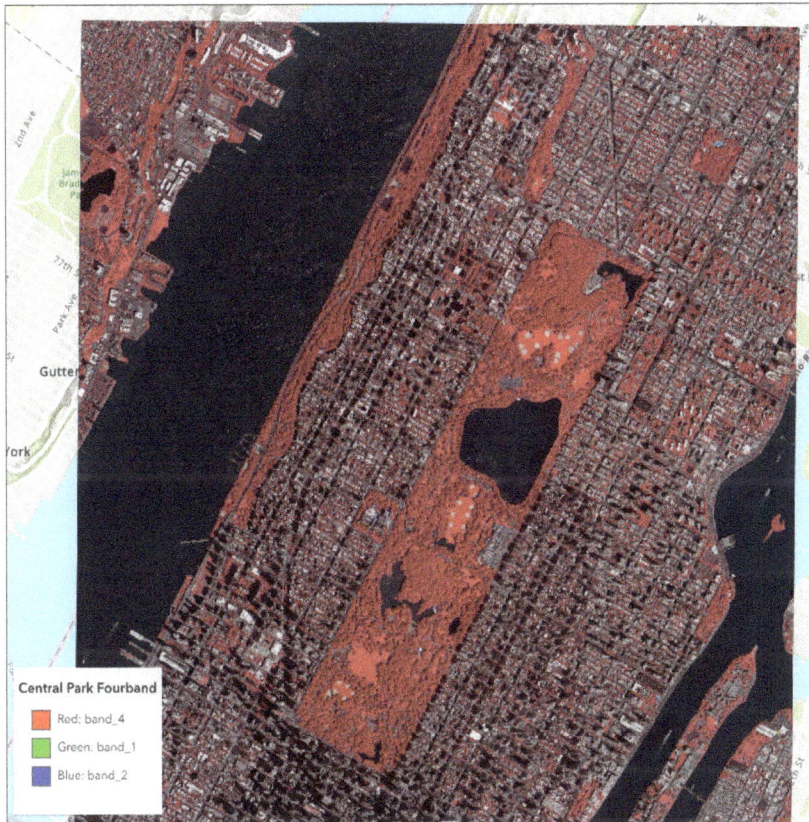

Now the imagery layer is displayed in the color infrared.

9. Zoom in to the map location to see the same area in this map that you previously viewed.

The imagery layer can be visualized and analyzed just as you did in ArcGIS Pro. Other options can also be used to improve the visual interpretation of the imagery layer.

One of the visualization options that is included is applying a raster function template to the imagery layer. You will learn more about how to create raster function templates in the next chapter, but there is one that has been created for you to use here.

Apply a custom raster function template

1. On the **Settings** toolbar, click **Processing templates** and note the templates that have been added to the imagery layer.

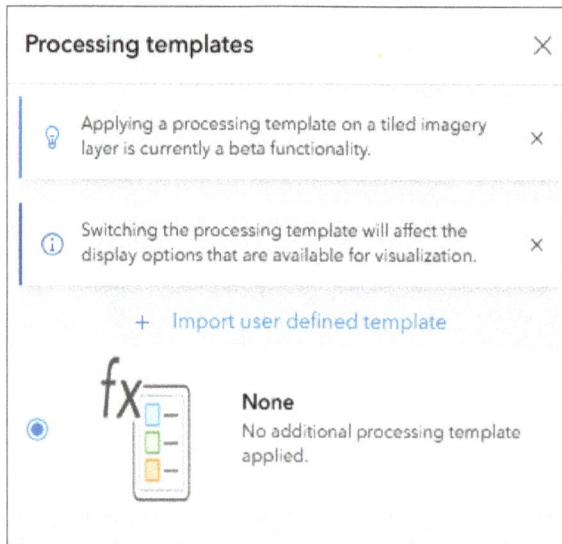

For this imagery layer, none are added, so you will import one.

2. Click **Import user defined template** to browse available templates.

3. Click **My content** and then click **ArcGIS Online.** Type owner: Top20Essential-SkillsForImageryAndRemoteSensing NDVI Scientific (Central Park_NYC).

4. Click the **Add** button (plus sign).

   This adds the processing template to your list.

5. In the **Processing templates** pane, make sure that **Custom-NDVI Scientific (Central Park-NYC)** is selected and click **Done**.

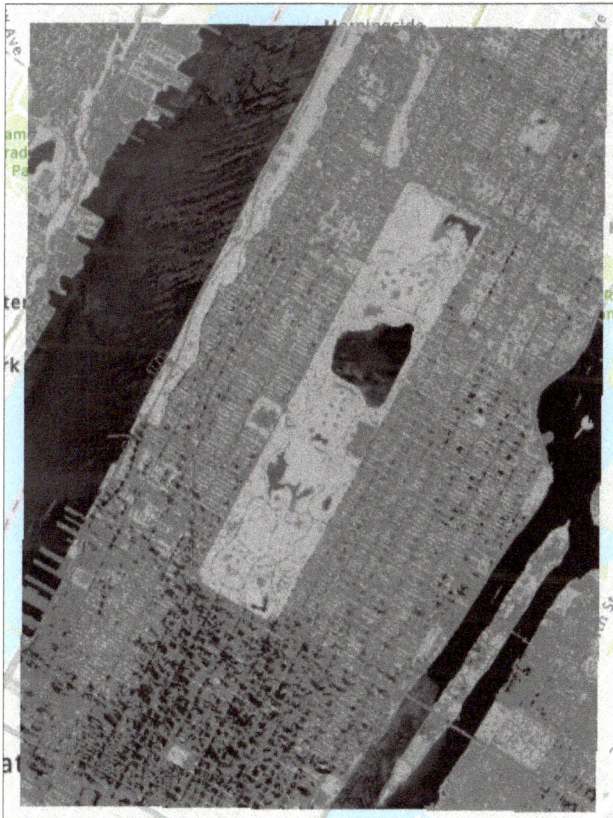

The NDVI raster function template has been applied to the imagery layer. Now you can see the raster band arithmetic function that has been applied to the imagery layer to see the imagery layer in a new way.

6. Zoom in to the map to take a closer look at the water body in the center of the park.

This raster function template uses band arithmetic to create a custom visualization, and you can see that some very different values are now visible that were not obvious in either the natural color or the color infrared band combinations.

7. Zoom out and consider the rest of the imagery layer.

You can see how healthy the vegetation is around the entire park in lighter shades of gray. The closer these hues approach pure white, the healthier and more vibrant the vegetation.

Sharing imagery layers to ArcGIS Online amplifies their ability to be used in your organization. You can share the imagery layers with other users and other applications allowing a common operating picture for shared analysis and understanding. ArcGIS Pro allows you to publish these imagery layers with all the custom settings of a mosaic dataset. Once published, the imagery layers can be added to maps or apps to continue the analysis.

Summary

In this chapter, you shared a mosaic dataset of NAIP imagery to ArcGIS Online from ArcGIS Pro. You applied several raster function templates to the image service as processing templates in ArcGIS Pro. Finally, you viewed this same image service in ArcGIS Online and applied the same processing templates there as well. These processes provide the building blocks of sharing images, particularly mosaic datasets, within your organization or to the public. From this starting point, you can begin analysis or use these shared imagery resources in other applications.

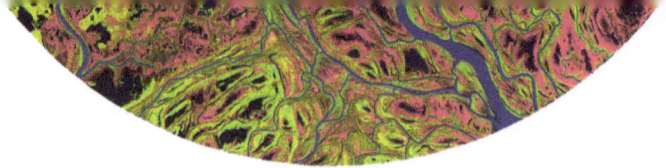

CHAPTER 19
Performing image analysis using ArcGIS Online

Jeff Swain and Thomas Humber

Objectives

- View an imagery layer in Map Viewer.
- Apply raster functions to analyze an imagery layer.
- Symbolize an imagery layer using a preset style.
- View an imagery layer in ArcGIS Pro.

Introduction

In this chapter, you'll learn how to analyze an imagery layer in ArcGIS Online Map Viewer using raster functions to create a new imagery layer based on your analysis results. Using ArcGIS Online, you'll see many familiar concepts after your experience using ArcGIS Pro. The important thing to remember is that the functionality to use the imagery layers and display them the way you want can occur in either application.

In this chapter, you'll add an imagery layer of Central Park in New York City to ArcGIS Online. Using this four-band imagery layer, you'll use a few raster functions in ArcGIS Online to analyze and identify areas of discrete, healthy vegetation growth. You'll take these results and apply a preset style to help show the results for decision-makers or others when they access your analysis results.

Tutorial 19-1: Analyze an imagery layer in ArcGIS Online

First, you'll open a new map in ArcGIS Online and add the imagery layer to the map to review the imagery that was created and then see all the data available to be analyzed.

> **Important:** *Your ArcGIS Online organization must have credits available to perform analysis on an imagery layer.*

You will consider the imagery layer properties and consider how to use the raster bands to create the desired analysis.

Open a browser to review item details

You will open a browser and then navigate to your ArcGIS Online organization to see the content created by your organization.

1. Open a web browser and go to arcgis.com.

2. Sign in with your credentials and click the **Content** tab.

   You will see all the content you have created in a section called **My content**. In addition to this section, you will see additional tabs for your favorites, groups, your organization, and ArcGIS Living Atlas. For this lesson, you will use an imagery layer of Central Park, in New York City.

3. At the top of the page, click the search bar and type owner: Top20EssentialSkills-ForImageryAndRemoteSensing Central Park NYC FourBand_Clip.

4. Under **Filters**, turn off the filter that limits your search to your organization and search all of ArcGIS Online.

5. On the **Central Park NYC FourBand_Clip** layer, click **Open in Map Viewer**.

Central Park NYC FourBand_Clip

Tiled imagery layer | Item updated: Jul 29, 2025

TO   Top20EssentialSkillsForImageryAndRemoteSensing | Open in Map Viewer   ...

This imagery layer has been clipped to show just the area of Central Park from a larger mosaic dataset of National Agriculture Imagery Program (NAIP) imagery. NAIP imagery is four-band imagery, including three visible bands and one near-infrared band. It also has a spatial resolution of 0.3 meters (0.98 feet). The imagery layer is currently visualized in the default **Natural color** visualization.

With this information, you will use this imagery layer to survey the health of the trees in the park. Because this imagery layer has four bands, you will use the near-infrared layer to create a Normalized Difference Vegetation Index (NDVI) of the area and locate the healthiest vegetation in the park.

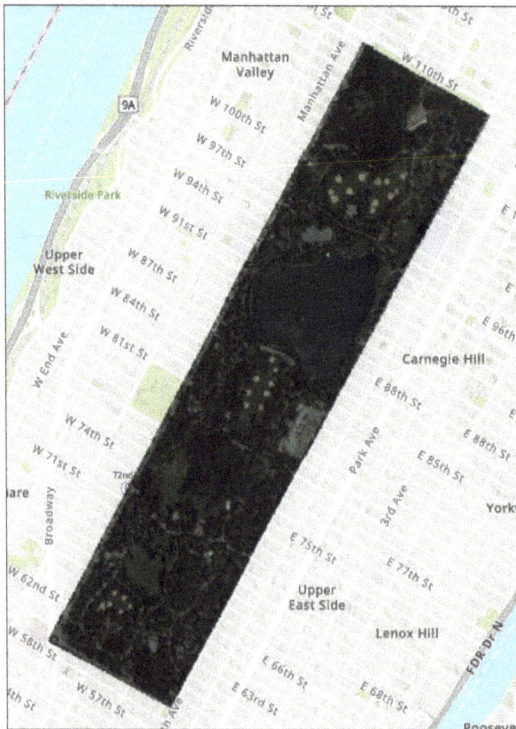

The imagery layer added to **Map Viewer** has all the pixel values and raster bands of the original source imagery and is ready to be used in your analysis. Before beginning any analysis, it is good practice to review the input data.

Modify the style of the imagery layer

In **Map Viewer**, you can alter the symbology of the imagery layer by modifying the style.

1. In the **Layers** pane, make sure that the **Central Park NYC FourBand_Clip** layer is selected.

2. On the right, on the **Settings** toolbar, click **Styles**.

   There will be two styles to choose from based on this imagery layer.

3. Under **RGB**, click **Style options**.

   You want to display a color infrared band combination to aid in your visual assessment and your analysis.

4. Under **RGB**, select the following bands for the appropriate color channels:
   - **Red**: Band_4
   - **Green**: Band_1
   - **Blue**: Band_2

5. Under **Stretch type**, change the type to **Percent clip**.

6. Click **Done**.

7. In the **Style** pane, click **Done**.

   Review the rendering in the map.

The imagery layer is displayed as color infrared, and the vegetation appears in vibrant shades of red.

Now that you have reviewed the imagery layer and viewed all four bands in the map, you are ready to start the analysis.

Perform an NDVI analysis on an imagery layer

1. In the **Layer** pane, next to the **Central Park NYC FourBand_Clip** layer, click the **Options** button and click **Zoom to layer**.

2. On the **Settings** toolbar, click **Analysis**.

    Here, you will see the analysis tools and raster functions to use in your analysis in Map Viewer.

3. At the top of the **Analysis** pane, click the **Raster Function** tab to see the available raster functions.

    > **Note:** Unlike raster functions in ArcGIS Pro, these raster functions will be used just like tools to generate a new imagery layer.

4. In the **Search By Name** field, type NDVI.

    There are two options available as raster functions for NDVI: **NDVI** and **NDVI Colorized**. You want to create an NDVI layer using scientific output values so you can apply preset thresholds to identify various levels of vegetation health.

5. In the **Raster Functions** pane search results, click **NDVI**.

    The **NDVI** raster function appears in the **Raster Functions** pane.

6. Apply the following settings for the **NDVI** function:
    *   **Raster:** Central Park NYC FourBand_Clip
    *   **Visible Band ID:** 1
    *   **Infrared Band ID:** 4
    *   Check the box for **Scientific Output**.
    *   **Output Name:** NDVI_CentralPark_<your initials>

7. Expand **Environment Settings**. Under **Processing Extent**, click **Display Extent**.

8. Once you have all the parameters set, click **Estimate credits** to confirm the inputs are correct and get an estimate of the required credits.

> **Note:** The estimated credits should be approximately 1. This process may take a few minutes.

9. Click **Run**.

The output is automatically added to the map after it finishes calculating the result. It appears as a single-band NDVI imagery layer.

Examine the NDVI analysis results

Because you selected the scientific output, the NDVI may look different from what you expect. This output measures vegetation health from −1 to 1. The value of −1 is for things such as sand, soil, pavement, and other nonvegetation objects. The closer a pixel is to 1, the healthier the vegetation is in that area.

1. On the left, on the **Contents** toolbar, click **Legend**.

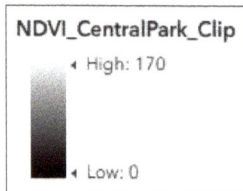

NDVI_CentralPark_Clip

◄ High: 170

◄ Low: 0

The max value for the layer is 0.7. You will alter the style to help determine the ends of the upper and lower values of the healthy vegetation. You know the upper end is 0.7, but now you need to find the lower threshold.

2. On the **Settings** toolbar, click the **Styles** button. Under **Select a style**, select **Classify** and click **Style options**.

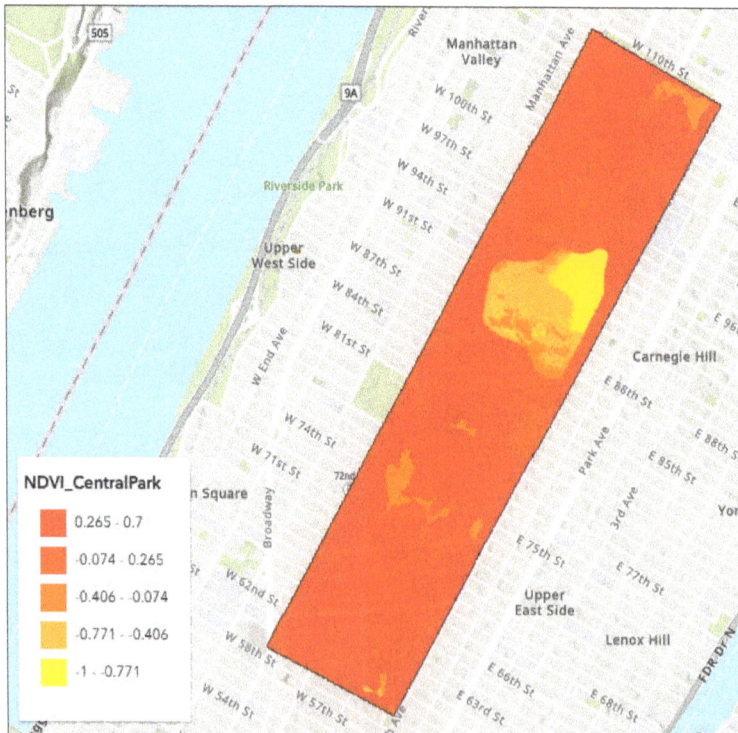

NDVI_CentralPark

- 0.265 - 0.7
- -0.074 - 0.265
- -0.406 - -0.074
- -0.771 - -0.406
- -1 - -0.771

You will see the imagery layer displayed in five classes and a **Yellow to Red** color ramp.

3. In the **Style options** pane, for **Number of classes**, change the value to 10 to see the highest threshold.

It is still difficult to see any differences, so you will change the color ramp to make the classes appear differently.

4. For **Color scheme**, click the **Edit** button (pencil).

5. In the **Color scheme** window, click the **Edit** button and then click **Red to Blue Diverging, Bright**.

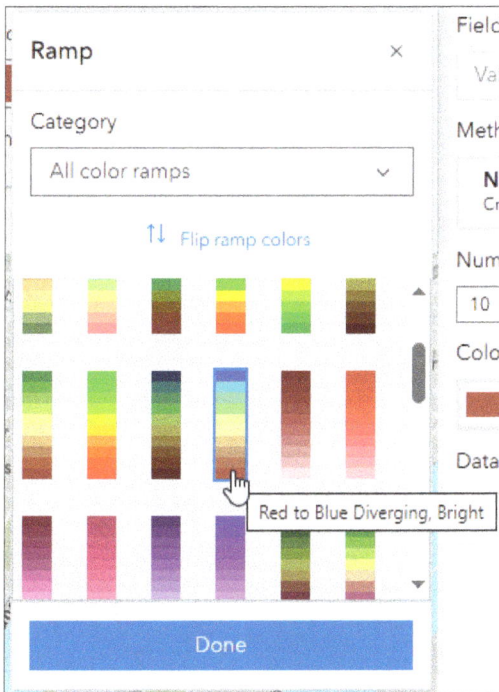

Tip: Hover over the color schemes to display the name.

The map changes color after your selection.

6. In the **Color scheme** window, click **Done**.

7. Close all open panels and view the map.

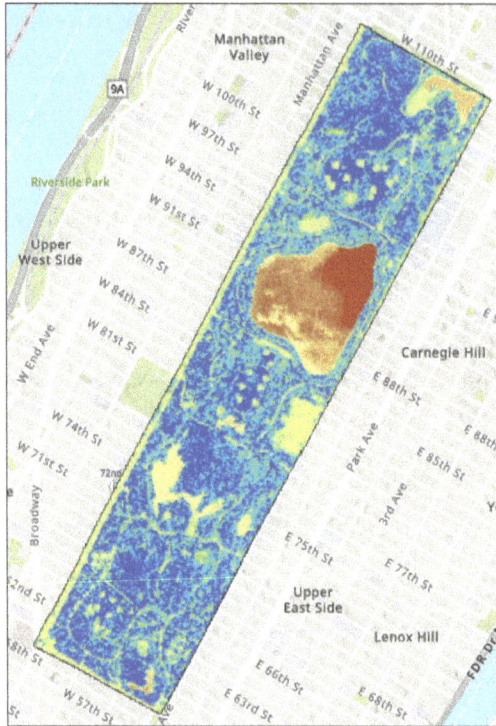

You can now see a greater difference as the upper limit is blue and clearly visible covering much of the imagery layer. Although this can approximate the healthy vegetation, you will make one more modification to see if you can find a more specific value for the healthiest.

8. Open the **Style options** again. Change the **Method** to **Quantile**.

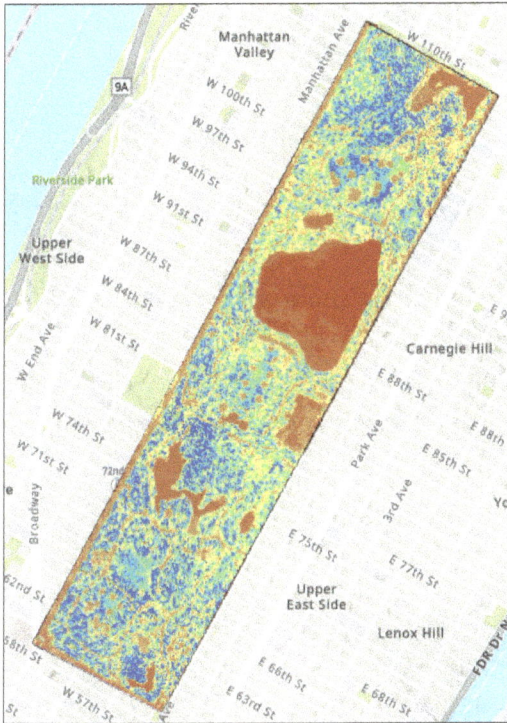

This further divides the classes and helps identify the upper value that will reveal the healthiest vegetation. The lower-end value of the top class is 0.5. You'll use this value, along with the maximum value of 0.7, to identify the healthiest vegetation according to the NDVI value.

Apply a threshold value using a raster function

Now that you have identified the upper and lower threshold values, 0.7 and 0.5 respectively, you can use these values to help isolate and show only the healthiest vegetation.

1. If necessary, zoom out to the full extent of the imagery layer.

2. Using the skills you learned earlier, open the **Greater Than Equal** raster function.

> **Remind me how:** On the **Settings** toolbar, click the **Analysis** button. Click the **Raster Function** tab and use the **Search By Name** field to find the tool.

3. In the **Greater Than Equal** pane, apply the following settings:
   - **Raster**: NDVI_CentralPark_<your initials>
   - **Raster2**: 0.5
   - **Output Name**: GreaterThanNDVI_CP_<your initials>

4. Expand **Environment settings**, and under **Processing extent**, click **Display extent**.

5. Once you have all the parameters set, click **Estimate credits** to confirm the inputs are correct and get an estimate of the cost in credits.

   Note: The estimated credits should be 1. The process may take a few minutes.

6. Click **Run**.

The areas that are greater than 0.5 will be displayed as a value of 1. Now you will complete the analysis by changing the visualization to create the desired rendering of the new imagery layer.

Note: The value of 0.5 is being used in this tutorial for example purposes. When isolating and identifying healthy vegetation, you'd probably use more rigorous analytic techniques along with additional information and resources. As a workflow illustration in ArcGIS Online, this value provides a good foundation for how to perform analysis.

Set the analysis result as a new layer and style

1. On the **Settings** toolbar, click the **Styles** button. For **Select Style**, select **Unique values** and then click **Style options**.

   The two values of 1 and 0 are displayed in two colors, red and gray. Now you will alter the style and create the preset style.

2. In the **Style options** pane, under **value**, click the gray square next to **0**. In the **Color scheme** window, click the **No color** button.

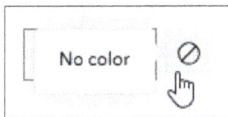

3. Click the red square next to **1** and change the color to a green of your choice.

You will see the new green imagery layer indicating the healthy vegetation overlaid on top of the other layer. To see the vegetation against the original image, you will turn off the visibility of the NDVI layer.

4. In the **Layers** pane, for the **NDVI_CentralPark** layer, click the **Hide** button (eye icon).

Now you can see the identified areas overlaid on the color infrared version of the original imagery layer.

As you zoom in to the map, you will see that only some of the trees are identified, showing which areas have the highest NDVI value. Although this isn't an absolute indicator of vegetation health, this visualization shows that there appear to be many healthy trees within the park.

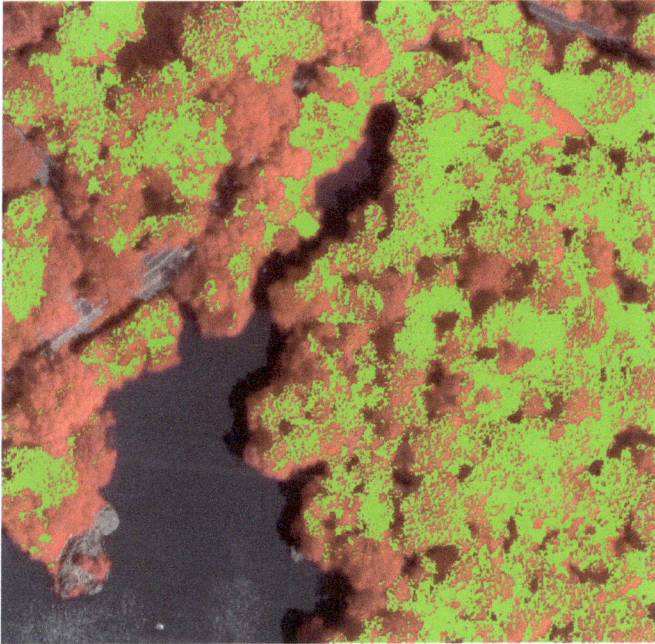

Create a preset style

Now that you have styled the imagery layer, you will save it as a preset style so that when other users add the imagery layer to a map, they can see it in the visualization you designed.

1. In the **Layers** pane, select the **GreaterThanNDVI_CP** layer.

2. In the **Style** pane, click **Style options**. Click **Register preset style**.

   Here you will name the preset style. This will label the option for all users of the layer.

3. In the **Register style** window, for **Style name**, type Healthy vegetation, click **Save**, and then click **Done**.

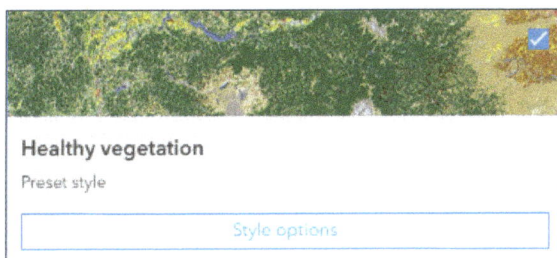

The preset style now appears above **Unique values** as an option to display the imagery layer. If you save the web map, this option will persist for this imagery layer in the map. Since you want to maintain the style for all users, you will save it with the layer.

4. In the **Layers** pane, make sure the **GreaterThanNDVI_CP** layer is selected. Click **Options** and then click **Save**.

   Now the preset style will persist for all those that add the layer.

5. On the **Contents** toolbar, click **Save and open** and then click **Save as**.

6. In the **Save Map** window, type the following:

   • **Title**: Healthy Vegetation Map of Central Park, NYC
   • **Tags**: Top20Imagery, Chapter 19, Central Park, NDVI Result, NYC
   • **Summary**: A map of Central Park, NYC, visualizing areas of healthy vegetation.

7. Click **Save**.

8. In the top-left corner of the screen, click the **Menu** button and click **Content**.

9. In **My Content**, for the **GreaterThanNDVI_CP** layer, click the **Options** button and then click **Open in Map Viewer**.

10. On the **Settings** toolbar, click the **Styles** tab.

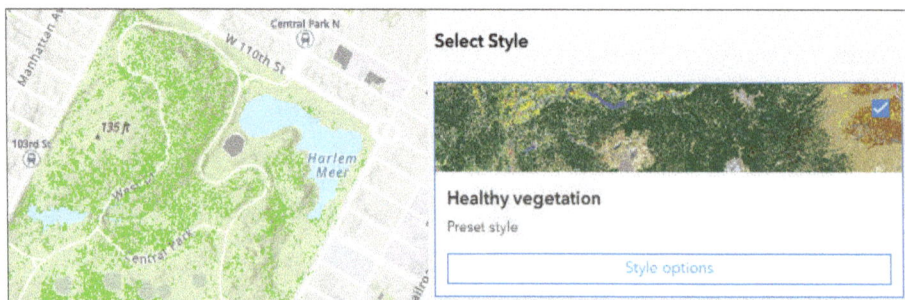

The layer still has your preset style.

You can add preset styles to imagery layers in Map Viewer, even those that have a processing template applied to them to create a new custom visualization. You can even see the custom visualization in ArcGIS Pro.

11. Open ArcGIS Pro.

12. In ArcGIS Pro, under **New Project**, click **Map**. Create a new map project titled Chapter19_CentralParkNDVI. Save it to a folder of your choice.

13. On the **Map** tab, in the **Layer** group, click **Add Data**.

14. In the **Add Data** window, on the left, under **Portal**, click **My Content**.

15. Search for and select the **GreaterThanNDVI_CP** imagery layer. Click **OK**.

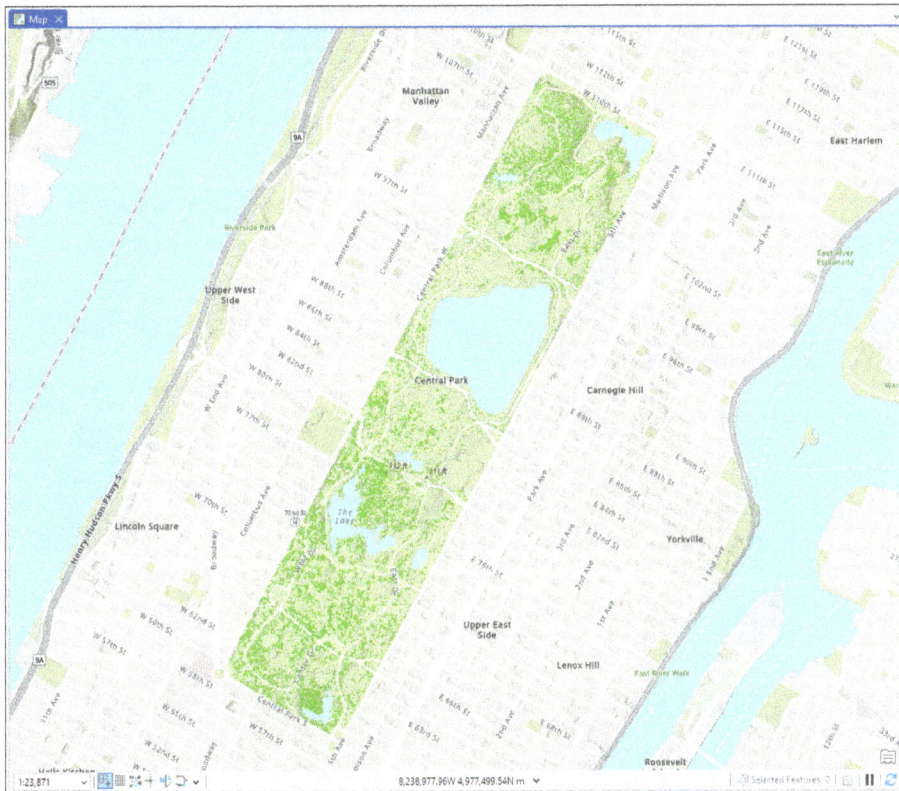

Analysis in ArcGIS Online can be used to augment your analysis in ArcGIS Pro and give you additional resources to complete your analysis. The style you create in ArcGIS Online can also be shared directly to ArcGIS Pro.

Take the next step

Image analysis, or raster analysis, in ArcGIS Online is available when you need additional resources to complete your analysis. In this tutorial, you used only a few raster functions on a small area to show the concept framework of analysis.

Using skills you've learned in several of these chapters, you can create function chains, build processing templates, and apply various raster functions or tools to help with your analysis. The generalized workflow remains the same. The possibilities of where your analysis goes is up to you and your imagination. The next step is to see where those avenues and resources take you!

Summary

In this chapter, you used several raster functions (NDVI and Greater Than Equal) to perform analysis on an image layer in ArcGIS Online. You examined the results of these analyses as part of a larger workflow to show healthy vegetation in the Central Park area. Finally, you used these results as a new layer as an overlay in Map Viewer to highlight areas of your isolated vegetation areas.

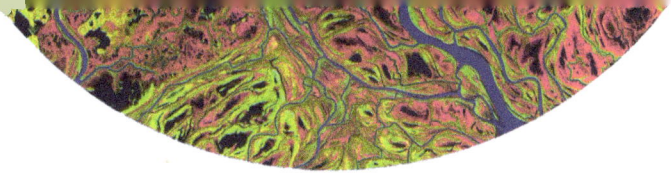

CHAPTER 20
Using your imagery skills in a decision support workflow

Objectives

- Introduce a decision support workflow.
- Identify relevant skills used throughout the processes of the decision support workflow.

Introduction

Imagery plays a critical role in decision support because it provides a real-world, up-to-date visual context for understanding situations, identifying patterns, and taking informed actions. Here's how imagery enables decision-making:

- **Situation awareness:** Imagery offers a clear real-time or recent view of conditions on the ground, helping decision-makers quickly grasp what is happening (for example, flooding extent, wildfire spread, urban growth).
- **Change detection and monitoring:** By comparing new imagery with historical data, stakeholders can detect change over time (for example, infrastructure damage after disasters, deforestation, shoreline shifts), enabling timely interventions.

- **Enhanced analysis and modeling:** Imagery serves as a foundation for creating derived products, including elevation models, land cover maps, and damage assessments, which feed into models and predictive analytics for planning and response.
- **Communication and collaboration:** Visual imagery makes complex data easier to interpret and share across agencies, improving coordination and ensuring that all stakeholders have a common operating picture.

As you've journeyed through this book, you've explored and learned various skills to manage, analyze, and share imagery using ArcGIS. Each chapter presents these skills as a unique aspect in its own solitary workflow. However, many of these skills work in concert with one another.

In this chapter, you'll see how these skills and techniques can work together to support a larger workflow. Unlike the other chapters—and because of the multifaceted complexity of this workflow—this chapter will outline a recommended framework and methodology for a larger problem.

When used together with the geographic approach—a way of thinking and problem-solving that integrates and organizes all relevant information through the crucial context of location—imagery provides a vital resource that can model scenarios and solutions, reveal patterns and phenomena otherwise hidden, and help identify trends that can enhance sound decision-making and situation awareness.

Tutorial 20-1: Use situation awareness and decision support

In this tutorial, you'll learn how skills outlined in previous chapters can operate together for a complete solution to a problem. The scenario is a hurricane strike on the eastern United States. A general workflow to understand situation awareness and work toward effective decision support for emergency operations follows three standard phases: **preparation**, **response**, and finally **mitigation and recovery**.

The East Coast of the United States has a dense population, as well as critical infrastructure, industry, and environmental areas of concern. Through each of these phases, analysts can use imagery as a cornerstone of their analysis and for making decisions.

PHASES OF DISASTER EMERGENCY MANAGEMENT

MITIGATION
Preventing future emergenies
or minimizing their effects

PREPAREDNESS
Planning for and reducing
the impact of disasters

RESPONSE
Protecting life and property
during a disaster

RECOVERY
Restoring the affected
area to normal operations

Figure 20-1. Imagery can be used to aid decision-making in each phase of the emergency management cycle.

Figure 20-2. Hurricanes are some of the most destructive events on the planet. Imagery can be used to help prepare populations, respond to damage and devastation, and recover from the impact of the storm.

Table 20-1. Spreadsheet shows which chapters address each step in the Preparation, Response, and Mitigation and Recovery phases of emergency management.

	2 Visualizing Imagery	3 Working with Raster Functions	4 Creating a Mosaic Dataset	5 Performing Radiometric Calibration	6 Georeferencing Imagery	7 Creating Photogrammetric Products
Preparation						
Identify and map critical infrastructure such as hospitals, shelters, roads, and utilities.	✓	✓	✓	✓	✓	✓
Plan and optimize evacuation and emergency response access routes.	✓	✓	✓		✓	✓
Allocate and position resources, such as where to place supplies and emergency personnel.		✓	✓			✓
Create public information systems using interactive web maps and dashboards.			✓			✓
Identify potential hazards such as flood zones, slope stability, or areas of impassibility.		✓	✓	✓		
Perform a risk assessment by combining hazard data with population, infrastructure, and land use.	✓	✓	✓	✓		
Identify vulnerable areas such as neighborhoods in floodplains.		✓	✓			
Conduct training simulations using real-world maps.	✓		✓	✓	✓	
Model scenarios to simulate possible disaster impacts.			✓			
Response						
Conduct real-time mapping and tracking using current weather maps, geospatial video, or other new collection means.	✓		✓	✓	✓	✓
Address ingress and egress routes mapping.	✓					✓
Perform damage assessment using satellite imagery, drones, or field data.		✓	✓	✓	✓	✓
Perform incident mapping.	✓		✓	✓		
Coordinate resource deployment.	✓		✓			
Integrate mobile apps for field reporting and data collection.	✓					
Mitigation and Recovery						
Analyze damage impacts.	✓	✓	✓	✓	✓	✓
Monitor reconstruction progress.	✓					
Conduct updated environmental assessment.	✓	✓	✓	✓		
Map and assess aid distribution.	✓					
Use imagery for long-term land use planning.		✓	✓	✓		

8	9	10	11	12	13	14	15	16	17	18	19
Performing Advanced Visual Analysis	Exploring Image Charts for Analysis	Working with Geospatial Video	Performing Change Detection	Performing Multispectral Classification	Using Deep Learning for Object Detection and Classification	Visualizing and Analyzing Scientific Multidimensional Raster Data	Exploring Hyperspectral Imagery and Spectral Analysis	Analyzing Synthetic Aperture Radar	Working with Image Services and Online Archives	Sharing imagery from ArcGIS Pro	Performing Image Analysis Using ArcGIS Online
				✓	✓				✓	✓	✓
✓		✓	✓	✓						✓	✓
									✓		
				✓	✓				✓	✓	✓
	✓			✓	✓		✓	✓			
✓	✓			✓	✓		✓	✓			
✓			✓	✓		✓		✓			
✓	✓	✓		✓	✓		✓	✓		✓	✓
										✓	
✓									✓		✓
✓			✓	✓	✓				✓	✓	✓
	✓	✓	✓	✓	✓			✓	✓	✓	✓
✓		✓	✓	✓	✓			✓	✓	✓	
✓		✓	✓						✓	✓	✓
✓	✓								✓	✓	✓
✓	✓	✓	✓	✓	✓	✓		✓	✓		✓
✓	✓	✓	✓		✓			✓	✓	✓	
✓	✓		✓		✓		✓	✓			
✓		✓	✓							✓	
	✓			✓	✓	✓	✓	✓			✓

Preparation

The **Preparation** phase is a key step in any decision support workflow. The goal of this phase is to reduce or avoid potential risks before a disaster occurs. Many steps can be taken before a hurricane arrives.

1. Identify and map critical infrastructure, such as hospitals, shelters, roads, and utilities.

    Effective disaster response relies on preparation, including getting foundational imagery ready in advance. Identifying critical infrastructure may involve tasks such as orthocorrection, radiometric calibration, traditional image classification, or applying deep learning workflows, with new imagery organized into mosaic datasets for efficient access and sharing. Ready-to-use and current imagery gives a reflection of real-world conditions and provides a vital head start for rapid planning and mapping in both expected and unforeseen events.

2. Plan and optimize evacuation and emergency response access routes.

    Planning evacuation routes after a hurricane's landfall requires using up-to-date imagery to create mosaic datasets, apply raster functions, and perform visual analysis skills. Geospatial video can play a key role in quickly gathering information, whereas sharing results through ArcGIS Online or ArcGIS Enterprise enhances accessibility and relevance. These preparations become critical once the storm hits, serving as a foundation for further analysis and response efforts.

3. Allocate and position resources, such as supplies and emergency personnel.

    Using insights from earlier steps, imagery can help identify optimal locations for staging resources and establishing emergency operation bases, such as new residential and commercial developments, large parking lots, or open spaces with appropriate access. Because GIS data may lag behind rapid community and infrastructure development, updated imagery is essential. Access to newly collected imagery, online collections, image services, and photogrammetric products equips planners with a powerful tool for positioning resources ahead of upcoming events.

4. Create public information systems using interactive web maps and dashboards.

As a storm nears landfall, updated imagery can support continued preparation in at-risk areas while also aiding public communication. Traditional tools, such as weather maps, radar, and forecasts, combined with new imagery, web maps, dashboards, and ArcGIS Hub sites, help convey critical information directly to the public. Skills, such as creating mosaic datasets, generating derived products from image classification, and using online archives, ensure the information shared is as current as possible.

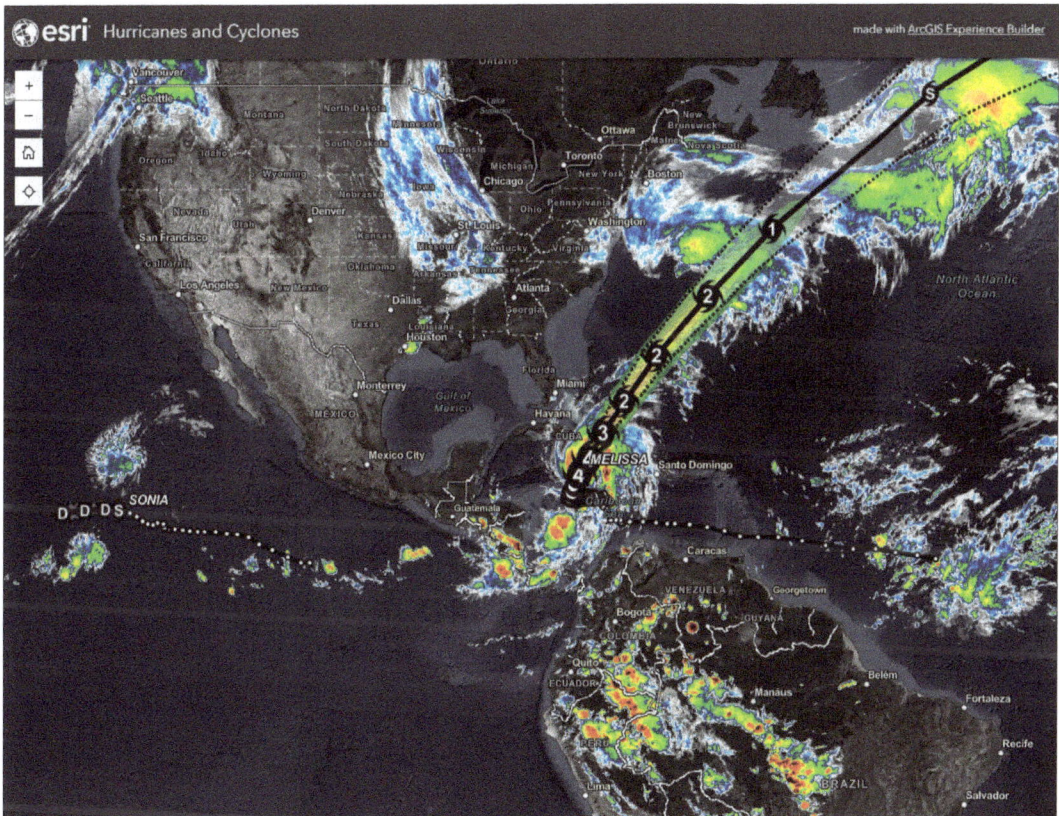

Figure 20-3. An example of an Esri Disaster Response Program public information app showing current weather conditions, active storm tracks, recent hurricanes, weather warnings, and colorized satellite imagery sourced from NOAA.

5. Identify potential hazards, such as flood zones, slope stability, or areas of impassibility.

    After identifying critical infrastructure, routes, and staging areas, the next steps focus on assessing hazards and risks. Image analysis techniques—such as using raster functions, classification, and deep learning using hyperspectral imagery (HSI) and synthetic aperture radar (SAR)—provide updated data and valuable resources for decision-makers. These processes help pinpoint potential hazard areas when used with elevation and other geospatial information.

6. Perform a risk assessment by combining hazard data with population, infrastructure, and land use.

    After identifying potential hazard areas, the next step is to evaluate their risk using the most current data from updated imagery. By combining GIS data, such as population, newly derived land use and land cover classifications, critical infrastructure layers, and updated route planning, you can categorize areas as low, medium, or high risk. Applying advanced visualization techniques to this combined analysis provides a comprehensive view of the regions most at risk.

7. Identify vulnerable areas, such as neighborhoods in floodplains.

    In this step, you'll use updated imagery and prior analysis to pinpoint vulnerable areas by incorporating multidimensional raster analysis for predictions based on historical events. Change detection using past and current imagery enhances the resources available for planning. Integrating these results with weather data, radar images, and storm tracks also helps prepare the public for potential storm impacts.

8. Conduct training simulations using updated maps.

    This step brings together more essential new skills than any other in the **Preparation** phase, serving as the bridge to the **Response** phase. Working with first responders, planners, and other stakeholders, you'll conduct training and simulations using new imagery, visualizations, image classification, change detection, geospatial video, and online imagery collections while sharing results in ArcGIS Online and ArcGIS Enterprise. This process may also require updating previous steps with new imagery or analyses to ensure readiness for the **Response** phase.

9. Model scenarios to simulate possible disaster impacts.

Because of the unpredictable nature of storms, updated training and simulations are used to model various scenarios, including different tracks, intensities, and directional changes. These simulations can incorporate National Oceanic and Atmospheric Administration (NOAA) and local forecasters' cones of uncertainty, combined with insights from updated imagery analysis and earlier preparation steps. All this information is often integrated into web maps, apps, or dashboards to provide comprehensive resources for those in need.

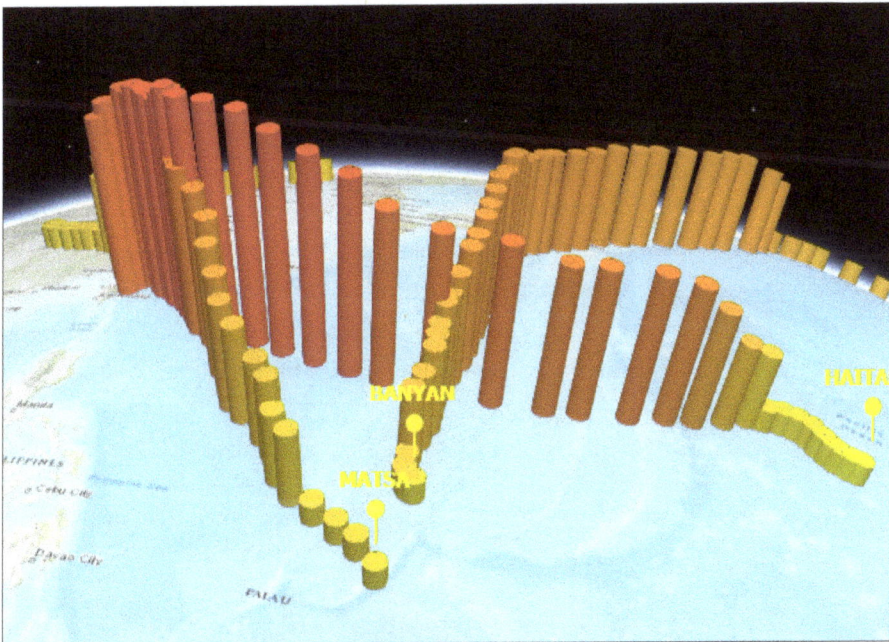

Figure 20-4. A 3D visualization of hurricane strength shows modeled storm tracks approaching land.

As you near the end of the **Preparation** phase of your decision support workflows, you can be confident that you've done everything possible to ready yourself, your team, and the affected community for the hurricane's arrival. Once the storm hits, the focus will shift to the **Response** phase.

Response

As the hurricane makes landfall, your response efforts must intensify. The **Response** phase focuses on delivering immediate aid, reducing damage, and saving lives during or immediately after the disaster. Timely image collections, such as geospatial video and SAR imagery, combined with modern information-sharing tools, such as web maps, experiences, and dashboards, have significantly enhanced the impact of emergency response. This is where imagery demonstrates its greatest value—often surpassing symbolic GIS maps in these real-world, critical, and rapidly changing situations.

1. Conduct real-time mapping and tracking using current weather maps, geospatial video, or other new collection means.

   Advancements in technology now provide faster-than-ever access to new imagery, whether collected with your own drones, CubeSats, or earth image constellations of satellites or from online resources that supplement real-time data. Skills honed during preparation and simulations are now applied in real situations, using updated weather data, radar, storm tracks, and newly captured imagery or geospatial video. This information forms the foundation for several upcoming steps in the **Response** phase.

Figure 20-5. NOAA imagery collected during Hurricane Helene 2024 of the Tampa, Florida, area. Imagery and web map available on the Esri Disaster Response Hub.

2. Address ingress and egress routes mapping.

   Now all your preparation starts to pay off! Newly collected imagery and geospatial video help assess whether planned routes are passable. Using tools such as SAR analysis, which can penetrate cloud cover, provides unmatched insight into on-the-ground conditions alongside your preplanned analyses. Are your routes blocked or open? Are there wide shoulders on some roads that allow the passage of emergency vehicles or turnouts where material and equipment can be staged? By combining initial datasets with newly collected data, you can deliver real-time or near-real-time route status updates to emergency responders.

3. Perform damage assessment using satellite imagery, drones, or field data.

   During hurricane damage assessments, you'll apply imagery and remote sensing skills, such as building mosaic datasets, updating the mosaics created in the **Preparation** phase, using raster functions, performing change detection, HSI and SAR analysis, and both traditional and deep learning classification techniques to identify damaged areas. As in many **Response** phase steps, imagery delivers updated, real-time information across large areas, offering a comprehensive view of the situation.

4. Conduct incident mapping.

   Once preliminary analyses are complete, the results—such as collapsed buildings, blocked roads, and other hazards identified from imagery, geospatial video, SAR, and deep learning—are shared with emergency responders. Ongoing monitoring and updates, including updating image classification and mosaic datasets, are performed as new information and storm impact reports arrive. Additional analysis may be needed depending on the storm's evolving status and track.

5. Coordinate resource deployment.

   Building on **Preparation**-phase data used for route evaluations and damage assessments, you can now apply resource staging and allocation information in this step. New imagery or geospatial video collected during or shortly after the storm helps pinpoint where aid is most needed as well as optimal staging and deployment locations depending on the situation on the ground. Using skills such as SAR analysis alongside traditional imagery from services and archives provides updated, real-time situation awareness as the storm weakens.

6. Integrate mobile apps for field reporting and data collection.

Throughout the **Response** phase, sharing imagery and analytic results through ArcGIS Online and ArcGIS Enterprise remains a continuous effort to support first responders, emergency personnel, and other stakeholders. Field data collected through tools such as ArcGIS Survey123 and ArcGIS Field Maps provides real-time updates that verify, enhance, and guide ongoing analyses. This integration of field reporting and imagery also helps direct collection resources to new locations for further inspection and assessment.

As the storm intensity diminishes and moves onward, your efforts must now move into the final phase of the decision support framework: the **Mitigation and Recovery** phase.

Mitigation and Recovery

Imagery and remote sensing analysis skills are integral to the final phase of the disaster response workflow. The **Mitigation and Recovery** phase focuses on restoring services, rebuilding infrastructure, and returning the region to normal—or even improved and more resilient—conditions.

1. Analyze damage impacts.

After the storm passes, cleanup and rebuilding efforts begin, supported by newly collected imagery prepared for analysis and integrated into existing mosaic datasets. Updates to prestorm and storm response products, including change detection through visual inspection, image classification comparisons, SAR analysis, and geospatial video, help reveal the storm's impact. These updated visuals and derived products can then be shared through various apps, dashboards, and ArcGIS StoryMaps stories to inform decision-makers, stakeholders, and the public.

2. Monitor reconstruction progress.

By applying visualization skills and using online resources, you can track the progress of recovery efforts. Techniques, such as visual change detection using swipe apps, deep learning image classification, SAR analysis for monitoring flood levels, and accessing online image catalogs, provide up-to-date insights. Sharing these results through dashboards, stories, and other apps keeps the public informed, builds trust, and supports resource allocation in the most affected areas.

3. Conduct updated environmental assessment.

   Post-storm recovery requires reassessing environmental damage, with many pre-storm studies and reports needing updates. Previous analysis workflows— such as creating mosaic datasets, visualizing imagery, and applying radiometric calibration or orthorectification—are repeated using recent imagery and derived products. Using traditional multispectral imagery (MSI), SAR, and HSI for change detection supports new environmental assessments addressing soil contamination, water quality, and areas of affected infrastructure.

4. Map and assess aid distribution.

   In addition to damage assessment and reconstruction, humanitarian aid distribution must be monitored and supported. Updated route analyses, new imagery, and geospatial video help evaluate how quickly aid reaches staging locations, whether recipients can access them, and if further analysis or route planning is needed. Once assessed, sharing the latest imagery and resources online with team members enhances the overall effectiveness of distribution efforts.

5. Use imagery for long-term land use planning.

   Although large storm impacts can't be fully mitigated, imagery and rasters can support intelligent planning for future events. Multidimensional rasters enable scientific analysis and long-term predictions, incorporating earlier analysis results as reference data. By applying analysis skills and sharing imagery, you can assist planners, decision-makers, and community leaders in recovery, rebuilding, change mapping, future preparedness, and improved resiliency.

   Once this final phase is complete, the cycle begins anew, with preparation, response, and recovery often blending seamlessly in a feedback loop. It's important to assess the many lessons learned during the process. Your enhanced imagery and remote sensing skills have not only helped you navigate the storm but also provided decision-makers and the public with valuable insights for the path ahead and aids for rebuilding resilient communities.

Summary

In this chapter, you examined the workflow and the three phases of a situation awareness and decision support strategy. Each step within these phases outlines workflows that showcase the key skills required to carry out the associated tasks. Imagery and remote sensing are most effective when integrated into a broader analytic framework, where timely image collection, strong analysis, and the synergy of ArcGIS enable you to maximize the full potential of imagery!

Workflow

Preparation

1. Identify and map critical infrastructure, such as hospitals, shelters, roads, and utilities.

2. Plan and optimize evacuation and emergency response access routes.

3. Allocate and position resources, such as supplies and emergency personnel.

4. Create public information systems using interactive web maps and dashboards.

5. Identify potential hazards, such as flood zones, slope stability, or areas of impassibility.

6. Perform a risk assessment by combining hazard data with population, infrastructure, and land use.

7. Identify vulnerable areas, such as neighborhoods in floodplains.

8. Conduct training simulations using real-world maps.

9. Model scenarios to simulate possible disaster impacts.

Response

1. Conduct real-time mapping and tracking using current weather maps, geospatial video, or other new collection means.

2. Address ingress and egress routes mapping.

3. Perform damage assessment using satellite imagery, drones, or field data.

4. Perform incident mapping.

5. Coordinate resource deployment.

6. Integrate mobile apps for field reporting and data collection.

Mitigation and Recovery

1. Analyze damage impacts.

2. Monitor reconstruction progress.

3. Conduct updated environmental assessment.

4. Map and assess aid distribution.

5. Use imagery for long-term land use planning.

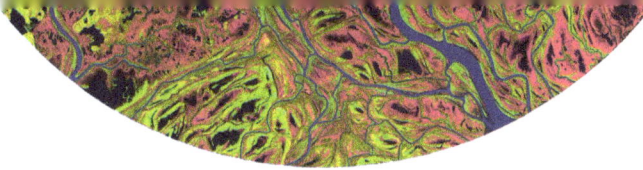

DATA SOURCE CREDITS

Chapter 2

Path 148 / Row 032 (30SEP2018) Level-1/Collection 2 (L1TP).

Earth Resources Observation and Science (EROS) Center, 2020. Landsat 8 – 9 Operational Land Imager / Thermal Infrared Sensor Level-1, Collection 2, US Geological Survey (USGS). https://doi.org/10.5066/P975CC9B.

Chapter 3

Path 111 / Row 075 (02FEB2019) Level-1/Collection 2 (L1TP).

Earth Resources Observation and Science (EROS) Center, 2020. Landsat 8 – 9 Operational Land Imager / Thermal Infrared Sensor Level-1, Collection 2, US Geological Survey (USGS). https://doi.org/10.5066/P975CC9B.

Path 111 / Row 075 (08MAR2020) Level-1/Collection 2 (L1TP).

Earth Resources Observation and Science (EROS) Center, 2020. Landsat 8 – 9 Operational Land Imager / Thermal Infrared Sensor Level-1, Collection 2, US Geological Survey (USGS). https://doi.org/10.5066/P975CC9B.

Chapter 4

Path 033 / Row 032 (03SEP2022) Level-2/Collection 2 (L1SP).

Earth Resources Observation and Science (EROS) Center, 2020. Landsat 8 – 9 Operational Land Imager / Thermal Infrared Sensor Level-2, Collection 2, US Geological Survey (USGS). https://doi.org/10.5066/P9oGBGM6.

Path 034 / Row 032 (26SEP2022) Level-2/Collection 2 (L1SP).

Earth Resources Observation and Science (EROS) Center, 2020. Landsat 8 – 9 Operational Land Imager / Thermal Infrared Sensor Level-2, Collection 2, US Geological Survey (USGS). https://doi.org/10.5066/P9oGBGM6.

Chapter 5

Path 233 / Row 078 (28MAR2025) Level-1/Collection 2 (L1TP).

Earth Resources Observation and Science (EROS) Center, 2020. Landsat 8 – 9 Operational Land Imager / Thermal Infrared Sensor Level-1, Collection 2, US Geological Survey (USGS). https://doi.org/10.5066/P975CC9B.

Path 233 / Row 078 (28MAR2025) Level-2/Collection 2 (L2SP).

Earth Resources Observation and Science (EROS) Center, 2020. Landsat 8 – 9 Operational Land Imager / Thermal Infrared Sensor Level-2, Collection 2, US Geological Survey (USGS). https://doi.org/10.5066/P9oGBGM6.

Chapter 6

Satellite Imagery © 2018 Vantor.

Chapter 7

Vexcel Imaging US Inc.

Chapter 8

Path 044 / Row 034 (20JUN2025) Level-2/Collection 2 (L2SP).

Earth Resources Observation and Science (EROS) Center, 2020. Landsat 8 – 9 Operational Land Imager/Thermal Infrared Sensor Level-2, Collection 2, US Geological Survey (USGS). https://doi.org/10.5066/P9oGBGM6.

Path 044 / Row 034 (19MAY2025) Leve-2/Collection 2 (L2SP).

Earth Resources Observation and Science (EROS) Center, 2020. Landsat 8 – 9 Operational Land Imager/Thermal Infrared Sensor Level-2, Collection 2, US Geological Survey (USGS). https://doi.org/10.5066/P9oGBGM6.

Earth Resources Observation and Science (EROS) Center, 2018. Digital Elevation – Shuttle Radar Topography Mission (SRTM) 1 Arc-Second Global: US Geological Survey (USGS). https://doi.org/10.5066/F7PR7TFT.

Satellite Imagery © 2018 Vantor.

Chapter 9

Path 195 / Row 026 (03APR2025) Level-1/Collection 2 (L1TP).

Earth Resources Observation and Science (EROS) Center, 2020. Landsat 8 – 9 Operational Land Imager/Thermal Infrared Sensor Level-1, Collection 2, US Geological Survey (USGS). https://doi.org/10.5066/P975CC9B.

Chapter 10

Esri Imagery and Remote Sensing team, Andrew Carey, Alex Posner. Location provided by City and County of Denver.

Chapter 11

Earth Resources Observation and Science (EROS) Center, 2024. Annual NLCD Collection 1 Science Products (ver. 1.1, June 2025): US Geological Survey (USGS). https://doi.org/10.5066/P94UXNTS.

Earth Resources Observation and Science (EROS) Center, 2024. Annual NLCD Collection 1 Science Products (ver. 1.1, June 2025): US Geological Survey (USGS). https://doi.org/10.5066/P94UXNTS.

Earth Resources Observation and Science (EROS) Center, 2020. Landsat Program, Collection 2, US Geological Survey (USGS). 1984–2020 (Landsat 5 TM, Landsat 7 ETM+, Landsat 8 OLI/TMS).

Chapter 12

Earth Resources Observation and Science (EROS) Center, 2018. National Agriculture Imagery Program (NAIP) US Department of Agriculture (USDA). https://doi.org/10.5066/F7QN651G.

NC OneMap (2019). North Carolina Department of Information Technology, Government Data Analytics Center, Center for Geographic Information and Analysis. Available at www.nconemap.gov.

Chapter 13

Copyright © 2021, Alaska Department of Fish and Game.

Copyright © 2021, Alaska Department of Fish and Game.

Chapter 14

NOAA's Precipitation Reconstruction over Land (PREC/L) data provided by the NOAA PSL, Boulder, Colorado, USA, from their website at https://psl.noaa.gov.

Chapter 15

Courtesy NASA/JPL-Caltech.

Chapter 16

Copernicus Sentinel data, 2018. European Space Agency (ESA).

Chapter 17

Copernicus Sentinel data. European Space Agency (ESA).

Chapter 18

Earth Resources Observation and Science (EROS) Center, 2018. National Agriculture Imagery Program (NAIP) US Department of Agriculture (USDA). https://doi.org/10.5066/F7QN651G.

Chapter 19

Earth Resources Observation and Science (EROS) Center, 2018. National Agriculture Imagery Program (NAIP) US Department of Agriculture (USDA). https://doi.org/10.5066/F7QN651G.

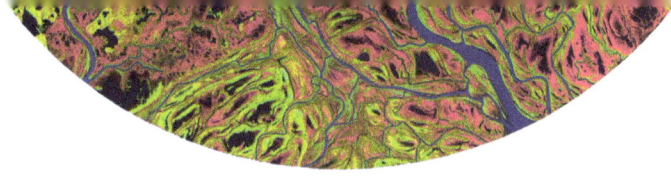

ABOUT ESRI PRESS

Esri Press is an American book publisher and part of Esri, the global leader in geographic information system (GIS) software, location intelligence, and mapping. Since 1969, Esri has supported customers with geographic science and geospatial analytics, what we call The Science of Where®. We take a geographic approach to problem-solving, brought to life by modern GIS technology, and are committed to using science and technology to build a sustainable world.

At Esri Press, our mission is to inform, inspire, and teach professionals, students, educators, and the public about GIS by developing print and digital publications. Our goal is to increase the adoption of ArcGIS and to support the vision and brand of Esri. We strive to be the leader in publishing great GIS books, and we are dedicated to improving the work and lives of our global community of users, authors, and colleagues.

Acquisitions

Stacy Krieg
Alycia Tornetta
Jenefer Shute
Katie Gezi

Product Engineering

Craig Carpenter
Maryam Mafuri

Editorial

Carolyn Schatz
Mark Henry
David Oberman

Production

Monica McGregor
Victoria Roberts

Sales & Marketing

Eric Kettunen
Sasha Gallardo
Beth Bauler

Contributors

Christian Harder
Matt Artz

Business

Catherine Ortiz
Jon Carter
Jason Childs

Related titles

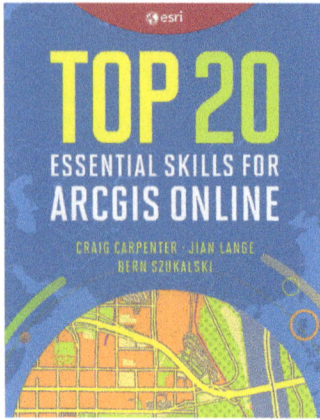

**Top 20 Essential Skills for
ArcGIS Online**

Craig Carpenter, Jian Lange
& Bern Szukalski

9781589487802

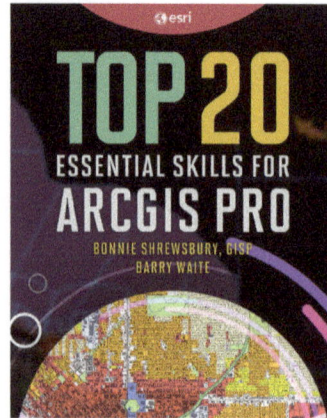

**Top 20 Essential Skills for
ArcGIS Pro**

Bonnie Shrewsbury & Barry Waite

9781589487505

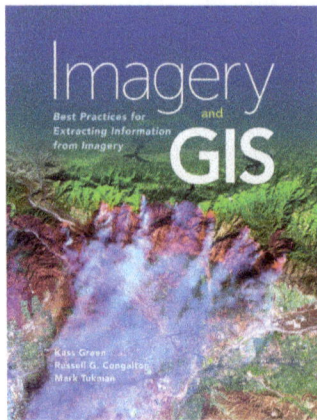

Imagery and GIS

Kass Green, Russell G. Congalton
& Mark Tukman

9781589484542

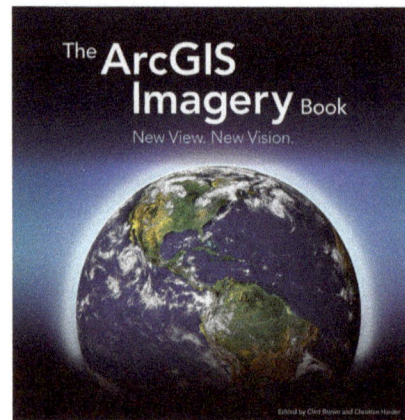

The ArcGIS Imagery Book

Clint Brown & Christian Harder (eds.)

9781589484627

For more information about Esri Press books and resources,
or to sign up for our newsletter, visit

esripress.com.